ECONOMIC INSTITUTIONS AND ENVIRONMENTAL POLICY

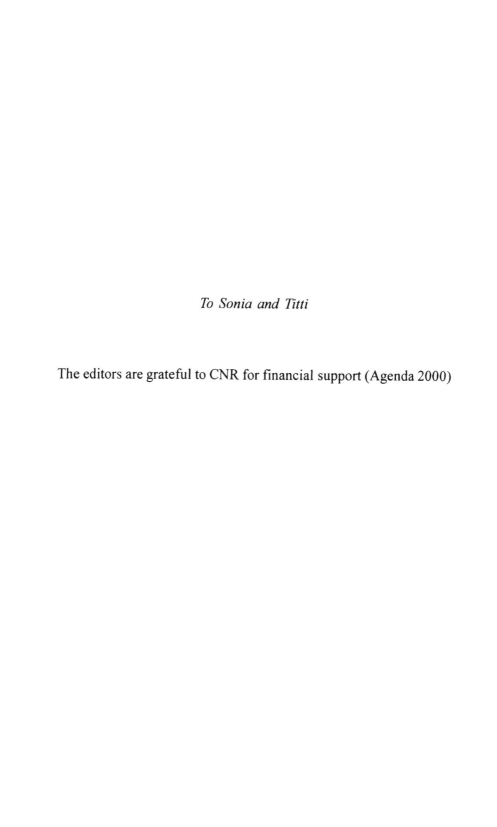

To Sonia and Titti

The editors are grateful to CNR for financial support (Agenda 2000)

Economic Institutions and Environmental Policy

Edited by

MAURIZIO FRANZINI
University "La Sapienza", Rome, Italy

ANTONIO NICITA
University of Siena, Italy

LONDON AND NEW YORK

First published 2002 by Ashgate Publishing

Reissued 2018 by Routledge
2 Park Square, Milton Park, Abingdon, Oxon OX14 4RN
711 Third Avenue, New York, NY 10017, USA

Routledge is an imprint of the Taylor & Francis Group, an informa business

Notice:
Product or corporate names may be trademarks or registered trademarks, and are used only for identification and explanation without intent to infringe.

Publisher's Note
The publisher has gone to great lengths to ensure the quality of this reprint but points out that some imperfections in the original copies may be apparent.

Disclaimer
The publisher has made every effort to trace copyright holders and welcomes correspondence from those they have been unable to contact.

A Library of Congress record exists under LC control number: 2001093280

ISBN 13: 978-1-138-70316-2 (hbk)
ISBN 13: 978-1-138-70319-3 (pbk)
ISBN 13: 978-1-315-20327-0 (ebk)

Contents

PART THREE: EVALUATION PROCESSES AND POLICY CHOICE

List of Contributors

Maurizio Franzini is Professor of Economic Policy at the University "La Sapienza", Rome.

Antonio Nicita is Associate Professor of Economic Policy at the University of Siena, Italy.

Simone Borghesi is Research Fellow at the Department of Economics, University of Siena, Italy.

Silvia Tiezzi is Assistant Professor of Economics at the University of Siena, Italy.

Marcello Basili is Associate Professor of Economics at the University of Siena, Italy.

Thomas P. Lyon is Associate Professor and Bank One Faculty Fellow in Business Economics and Public Policy in the Kelley School of Business at Indiana University.

John W. Maxwell is Associate Professor of Business Economics and Public Policy in the Kelley School of Business at Indiana University.

Massimo Di Matteo is Professor of Economics at the University of Siena, Italy.

Salvatore Bimonte is Assistant Professor of Economics at the University of Siena, Italy.

Sergio Ulgiati is Assistant Professor of Chemistry at the University of Siena, Italy.

PREFACE

1. Economic Institutions and Environmental Policy:
An Introduction

Maurizio Franzini and Antonio Nicita

The influence of economists on environmental policy has increased significantly in recent years. In many countries, economic instruments and evaluation techniques proposed by economists have had a broad application.

However, this ascendency of environmental economics has been challenged by new phenomena of environmental degradation or by the worsening of old ones, often on a global scale, which have become a serious threat to humankind. The new aspects of environmental degradation pose a serious challenge to environmental economists, despite the fact that the origin of these phenomena presumably depends more on the characteristics than on the intensity of economic growth. It is time to draw a balance of the real contribution that economists can make towards sustainable development (Hahn, 1999).

Economists like Dasgupta (1990) argued that environmental economics has achieved the peak of its theoretical research, so that its only contribution at this point is to monitor policy implementation (such as implementation of Pigouvian taxes, privatisation of common property, energy saving).

This book challenges this view and argues that the effectiveness of environmental policy depends on the role played by economic institutions in affording and implementing policy objectives. Theoretical research into an important issue of environmental economics, namely the relationship between market failures and 'institutional failures', has been quite neglected by economists.

In order to assess the two-way traffic between economic institutions and environmental policy, two questions should be asked:

1. how do economic institutions evaluate and implement environmental policies?
2. to what extent do 'institutional failures' affect the effectiveness of environmental policies aimed at solving market failures?

To answer these questions means to investigate the role played by institutions in affecting and implementing environmental policy. Institutions, in the wider sense, are typically shaped by a complex set of economic and political interests, culture and values which require detailed interdisciplinary analysis.

Beyond the notion of externality

The traditional way in which economists approached environmental problems was essentially based on the notion of externality, i.e. on the notion of market failure in dealing with resources which typically did not pass through market exchange and hence through a price-based system.

This traditional approach turned out to be unsatisfactory for at least two reasons:

1. environmental problems often go beyond the notion of externality, in the sense that they would probably arise even in a world of well-functioning markets;
2. traditional economic instruments designed to solve environmental externalities are often ineffective or fail to fulfill the relevant policy aims.

Because both these questions refer to the role played by economic institutions in dealing with environmental policies, they require theoretical analysis which oversteps the boundaries of the market paradigm and suggests a wider theoretical toolkit which takes interactions between alternative institutions into account.

The solution of an environmental externality is not an isolated decisional process between two stylised agents, the 'polluter' and the victim, but rather a complex *social* process involving the definition of property rights, equity issues, administrative costs, individual preferences, and so on. When we depict an environmental externality as a stylised non-market transaction between two individuals or collective-type agents, we are neglecting a wide range of *second-order decisions* which may heavily affect the ultimate outcome of that stylised transaction.

According to Sunstein and Ullman-Margalit (2000, p.187), second-order decisions are "decisions about the appropriate strategy for reducing the problems associated with making a first-order decision".

Let us think, for instance, about application of a Pigouvian tax by the government of country X, in order to solve a given environmental problem.

The evaluation process and political process which determine the eventual tax may be complicated and time-consuming. For example, a second-order decision may be any step taken by the victim's association to exert social or political pressure on the government to levy an environmental tax on polluters. However, before acting as an association, the victims have been considered how to promote a political association of victims against the polluters, and so on. Moreover, when the tax is levied by the government, an environmental agency presumably has to monitor whether the tax level is appropriate to induce polluters to diminish their pollution level. In this case, the environmental agency has to solve a second-order decision about how to monitor effectively, and so on.

It is therefore easy to imagine a wide range of second-order decisions, besides the first-order decision calculated to solve a given environmental externality. The efficiency of second-order solutions in lessening the burden of subsequent decisions depends on the institutional order in which the decisions are made.

In order to properly assess the degree of success of any environmental policy, we therefore also need to evaluate the effectiveness of second–order decisions and the respective institutional order. This conclusion has largely been neglected in economic theory and environmental problems have often been analysed as first-order decision problems in the market framework.

Consequently, the notion of externality has primarily been defined in terms of (market) (in)efficiency and the solution of environmental problems has been identified with the problem of promoting policies aimed at replicating a market system.

The emergence of second-order decisions therefore requires however an extension of the traditional economic approach to environmental problems in order to take into account the wide range of 'institutional failures' which might be encountered in the adoption of environmental policies.

There are at least two different issues. The first is detection of a series of institutional failures which go beyond the traditional market failures. The second is analysis of the 'externalities' produced by second-order decisions on first–order decisions, i.e. analysis of the effectiveness of economic instruments in dealing with environmental problems, when some institutional failures persist over time.

With reference to the first issue, a central point is the role played by agents' preferences in producing, defining and solving an environmental externality. These preferences, far from being well-informed, endogenous and customised, are assumed by economists to be given and constant in time. The process by which preferences are revealed is another important issue here, with respect to the definition and solution of an environmental problem.

Another problem related to the second issue is the existence of serious obstacles to the effectiveness of environmental policies, such as slow policy implementation and the 'inertia' of policy decisions.

How 'inertia' affects Pigouvian policies

On the whole, traditional environmental economics seems to take a simplified (and often unsatisfactory) approach to the two major questions: the selection of environmental objectives and how to implement them.

With regard to objectives, a first challenge is whether to continue to employ the notion of efficiency as a main guide for economic theories; with regard to means, it is necessary to see whether the alternatives of market solutions (in the tradition of Coase) and central intervention (according to the approach of Pigou) are still the only instruments available.

The emphasis on efficiency as an essential criterion of evaluation is sometimes excessive, often being difficult to apply because of the many unconventional (and not always easily measured) "costs" and "benefits".

These considerations raise the need for a more articulated basis of evaluation and the need to include elements, such as equity (between and within generations), that are often ignored, in the calculations.

It has recently become evident that it is necessary to go beyond the idea that the task of environmental economics is merely to implement the solution appropriate to a perfectly competitive market.

Environmental economists often propose an over-simplified view of the means to attain the objectives, whether Pigouvian or Coasian. The Pigouvian approach takes an ingenuous view of public intervention, ignoring the difficulties of policy implementation and the Coasian approach assumes that it is possible to create the conditions that enable the market to reach an efficient configuration, absorbing all the environmental externalities generated.

To illustrate this more clearly, let us think of the slowness or 'inertia' which characterises many policy decisions. "Doing nothing" or delaying decisions is often a serious problem in this field, practically ignored by economists. This is somehow paradoxical: environmental economics does not address a central point in defining objectives ("act quickly" could be an objective) or in selecting instruments (instruments that can be applied quickly should also be favoured). Ignoring the effectiveness of the instruments selected has earned economists the reputation of being unrealistic or even superfluous.

Of course, "actuation" time can be introduced as a criterion of efficiency, but if it remains a merely "nominal" question, it will certainly not help in solving the environmental problem at stake. Our idea is that the extraordinary slowness with which important environmental decisions are made is evidence of "institutional failure" to solve environmental problems.

Institutional failures and the Coase Theorem

Institutional failures may reverse some of the main conclusions of the Coasian perspective. To clarify this point, it is useful to have another look at the approach of Coase, the main pillar of which is the controversial theorem of Coase. Paradoxically, this theorem can help us to understand where the dangers lie for distributional questions, and hence the causes of operational and decisional inertia in the environmental field.

According to the theorem, in the presence of externalities: "if property rights are well defined and if market transactions are costless, then possessors of those rights will agree, by bargaining, on an allocation that is both Pareto optimal and independent of any prior assignment of property rights", or in the formulation of Cooter (1987): "... the initial allocation of legal entitlements does not matter from an efficiency perspective so long as the transaction costs of exchange are nil."

The Coase approach therefore enables environmental externalities to be explained in terms of transaction costs and the definition of rights. The "definition" of rights for any possible action that could have an indirect or external effect, seems to be a central element, since without rights there would be no market transaction.[1] Once the big question of the definition of rights has been solved, Coase's thesis seems to apply to any *Pareto-relevant* transaction, blocked by the existence of transaction costs.[2] The immediate consequence is that in the analysis of Coase, the economic

---

[1] As underlined by Coase (1960, p.8): "... it is necessary to know whether the damaging business is liable or not for damage caused, since without the establishment of this initial delimitation of rights there can be no market transaction to transfer and recombine them". Cooter (1987) also pointed out that "ensuring the efficiency of the law is a matter of ensuring the existence of perfectly competitive markets".

[2] As Stephen (1988, p. 27) writes: "The Coase theorem is in fact an application of a general principle that exchange only takes place when it is beneficial to both parties."

centrality of transaction costs is not a specific characteristic of the externality, but an aspect common to any market transaction.

In the trading of rights, if the transaction costs do not exceed the advantages expected from the transaction, the rights will go to the subject capable of maximising the value. This exchange, however, is not the internalisation of an externality, but more often a reallocation of resources.

This enables us to distinguish the process of definition of rights from that of reallocation of resources through market transactions. What happens specifically when the definition of rights is an infinitely costly process? For example, in the case of transnational commons, does the absence of rights impose an insuperable distributional constraint on Coasian exchange to restore efficiency? Are we doomed to the inertia of Coasian exchange and hence to the inertia of the environmental externality?

It seems clear that in all these cases the "inertia" of the Coasian process is due to transaction costs for the definition of rights rather to transaction costs of rights already defined. Coasian analysis therefore cannot answer the questions posed above, because its aim is only to identify the conditions that enable the market to function well. Coasian analysis has thus the limitation of tailoring economic analysis to the role of market exchange. To overcome this weakness, the market can no longer be assumed to be the only frame of reference. We need to look at the whole economic, political and environmental context, bearing in mind that the 'market' plays an important but not comprehensive role in dealing with externalities. Exclusive emphasis on market function means favouring some objectives and ignoring others, so as that theoretical analysis is relegated to such a level of abstraction as to be incompatible with effective policy.

Markets do not function perfectly and market failure is more often the rule rather than the exception. A perspective that recognizes broader institutional failure as far as the environment is concerned is therefore needed.

Many cases of institutional failure, ranging from the problem of defining property rights to environmental resources, to the problem of actuating environmental policy in a reasonable time, not only fail to solve the externality, but create inertia in decision-making, aggravating environmental damage and increasing its irreversibility.

Among the underlying causes of this inertia in solving the environmental externality or implementing policy is failure to reach agreement between the parties, which in turn is almost always a matter of distributional conflicts. In economic analysis and more specifically environmental economics, distributional questions have been almost completely ignored. Efficiency, defined in the above unsatisfactory way, tends to be considered a priority and is somehow depicted as independent of distributional questions. The idea underlying this attitude is that if the pie grows, it is easier to find a way to give a bigger slice to everyone. The real situation is quite different: when there are obstacles to redistribution policy, distributional questions can make it impossible to reach any effective agreement on the preconditions for implementing efficient responses to environmental degradation.

Distributional conflicts and efficiency

Distributional conflicts not only challenge the Coasian framework, but also entail the effectiveness of Pigouvian policy solutions. In the latter case, in recommending solutions or deciding the quantity of resources to allocate for environmental protection, the regulating agency should try to reach decisions quickly, but the reactions of parties who fear to be damaged may block any initiative.

The choice of the regulating agency is influenced by the specific institutional context. As Yandle (1999) remarked: "if we are to focus on choosing, the normative assumption of welfare maximization [which characterizes the Pigouvian tradition] must be replaced with a positive analysis of political choice", since "in a political system where votes determine outcomes, special interest groups have operational incentives to seek favours or rents in the resulting political economy." Yandle offers a good review of the contributions that emphasize the "political" efficiency of environmental choices in industrialized countries, from the traditional analyses of, Tullock (1962) and Olson (1965). According to Yandle, most political choices can be explained on the basis of public choice theory, rather than on the basis of Pigouvian or Coasian prescriptions.

Transaction costs must therefore also include costs of influence (Milgrom and Roberts, 1992) and rent-seeking, since these costs may prevent the definition of rights (and hence absorption of the externality) and trading of rights once they are defined (and hence attainment of efficient allocation).

As we gradually eliminate the Coasian hypotheses of completeness of rights (which implies zero transaction costs in the definition of rights) and completeness of markets (which implies zero transaction costs in negotiating defined rights), we find that the classical instruments for comparing different political solutions (including the criterion of Pareto efficiency) lose any plain meaning and negotiation between economic interests mingles with phenomena of political organization and representation of private interests and with strategies for "influencing" political decision-makers. As a consequence, both decentralised market solutions *à la* Coase and centralised solutions *à la* Pigou both encounter serious problems.

As Kapp (1963) observed, the economy is never completely free of influences of a political and ethical character, since conflicts of interest and coercive elements creep into the political process, influencing the evaluation of social benefits (and costs) and the determination of social priorities.

To compare transaction costs of two different institutions governing a given transaction, single decision-makers and the regulating agency must have access to correct information at a negligible cost. The cost of acquiring and processing the information necessary to evaluate alternative transaction costs is in turn a "transaction cost", the size of which depends on the institutional context in which the choice is made, and so forth.

It therefore seems increasingly important to investigate the double causative relationship between evaluation processes and institution structure: the former are the result of the institutional context in which they occur; the institutions in turn arise on the basis of comparative evaluation of the respective costs sustained in governing transactions. What emerges clearly is that only by adopting a more complex institutional approach can progress be made in the direction briefly indicated here. This approach should take into account the complex relations existing between evaluation and institutions and should place the role of the market in a more complex institutional context, so as to overcome the overdependence shown by the analyses of environmental economists on this nevertheless important institution.

The plan of the book

The essays in this volume examine the questions of evaluation processes and enforcement with regard to environmental externalities, stressing the role of institutional failures in dealing with the environment.

The essays are divided into three sections: 1. the complexity of evaluation processes; 2. the complexity of institutional arrangements; 3. evaluation processes and political choice.

Borghesi discusses how the inertia of environmental policies depends on the fact that policy has the sole aim of economic growth, which by generating more resources for everyone, might at any time implement the optimal resorption of environmental externalities within and between generations. Lyon and Maxwell, Nicita and Di Matteo show that when the Coasian market and Pigouvian regulator fail, endogenous enforcement mechanisms with high efficiency in terms of environmental protection and fast decisions can be reached by political negotiation and coordination of environmental policy with other economic policy. As Borghesi and Tiezzi, Bimonte and Ulgiati, and Basili point out, the efficiency of environmental policy is based on an appropriate process of ex-ante evaluation of relevant variables, which in turn can by corrected dynamically by ex-post evaluation of the same variables, as analysis is gradually refined, agents' preferences are revealed and scientific information becomes more reliable.

As Franzini observes in the concluding note, evaluations and institutions are sides of the same coin. This means that the complexity of the institutional order in which transactions occur, brings to light a variety of alternative institutional solutions to problems of environmental externality related to the existing institutional regime, avoiding the traditional dichotomy between "decentralized" market solutions and "integrated" organizational solutions.

REFERENCES

Coase R. (1960) "The Problem of Social Cost", *Journal of Law and Economics*, 3:1-44.

Cooter (1987) "The Coase Theorem" in *The New Palgrave*, MacMillan.

Dasgupta P. (1990) "The Environment as a Commodity", *The Oxford Review of Economic Policy*, Vol. 6, Issue 1, pp. 51-67.

Hahn R. (1999) "The Impact of Economics on Environmental Policy", *AEI-Brookings Research Paper* n. 99-4.

Kapp W. (1963) "Social Costs and Social Benefits. A contribution to Normative Economics, in E.V. Beckerath and H. Giersch (Eds), *Probleme der normativen Ökonomik und der wirtschaftspolitischen Beratung. Verein für Socialpolitik*, Duncker&Humblot, Berlin

Milgrom P. and Roberts J. (1992) *Economics, Organization and Management*, Prentice Hall International Editions.

Olson M. (1965) *The Logic of Collective Action*, Cambridge, CUP.

Stephen (1988) *The Economics of the Law*, Wheatsheaf Books.

Sunstein C. R., Ullman-Margalit E. (2000) "Second Order Decisions" in Cass R. Sunstein (Eds) *Behavioral Law & Economics*, Cambridge University Press, pp. 187-208.

Tullock G. (1962) "The Welfare Cost of Tariffs, Monopolies, and Theft", *Western Economic Journal* 5 (June): 224-232.

Yandle B. (1999) "Public Choice at the Intersection of Environmental Law and Economics" in *European Journal of Law and Economics*, 8:5-27.

PART ONE
THE COMPLEXITY OF
EVALUATION PROCESSES

2. Welfare Indices and Environmental Accounting:
A Critical Survey

Simone Borghesi and Silvia Tiezzi[*]

Introduction[1]

One of the most controversial issues in National Income Accounting is the research of an appropriate welfare indicator. The Gross National Product (GNP) is often used for this purpose. However, as El Serafy and Lutz (1989, p.1) point out, the concept of welfare "is much broader than a monetary measure of income", since it encompasses many aspects of human well-being that cannot be measured in monetary terms.

One of the main shortcomings of GNP is that it does not adequately reflect the depletion and degradation of natural resources, which makes it diverge from a true measure of income. For this reason the Net National Product (NNP), which takes capital depreciation into account, has been suggested as a more suitable welfare measure. As it is well known, NNP is defined as follows:

NNP = C + I + (X-M)

where C = consumption, I = net investments, X = exports and M = imports.

If we assume that welfare depends on consumption possibilities, then the first term on the right-hand side can be interpreted as current well-being from production today, and the remaining two terms as future

[*] We thank Geoffrey Heal and Kirk Hamilton for very useful discussion on this subject. All remaining errors are only ours.

[1] Although this paper was jointly written, Simone Borghesi takes responsibility for section 2 and Silvia Tiezzi for section 4.

consumption possibilities from current investment activities. More precisely, net investments represent the increase in future production capacity of the economy, while net exports (X-M) imply an accumulation of claims on other countries that will eventually lead to a larger amount of imported consumption goods in the future.

As Mäler (1996, p.3) has pointed out, this definition of NNP is still a rather narrow notion of welfare since it includes only consumption goods that are bought and sold on the market. In fact, current welfare is also influenced by non-market goods and services such as environmental amenities. Similarly, future wellbeing depends on variation of assets that are not transacted on the market, such as depletion of exhaustible resources and net changes in the stock of renewable resources. These considerations raise the following issue: how can we adjust national accounts to reflect the economic depreciation of natural resources? In this regard, we can distinguish two main approaches that have tried to answer this question. On one hand, economic theorists have used optimal control theory to derive a correct measure of environmental degradation in mathematical terms. On the other, national accountants have extended the System of National Accounts (SNA) in the form of satellite accounts, the main result being the System of Economic and Environmental Accounts (SEEA) proposed by the United Nations in 1993.

The difficulty in translating the adjustments suggested by economic theory into an accounting tool can probably explain why the theoretical approach has often received little attention at empirical level, so that "the various groups proposing answers are not communicating with each other" (Mäler 1996, p.4).

The aim of the present paper is to make a contribution that can enhance communication between these different groups. For this purpose, we will first try to explain the arguments put forward by economic theory to arrive at a correct welfare measure and then investigate how the adjustments emerging from the theoretical analysis can actually be computed. We will not examine, instead, the existing macroeconomic accounting tools that could be used as a framework to include these adjustments. This is because a proper treatment of the integration between environmental and economic accounts would require a detailed analysis that is beyond the scope of the present work.

The structure of the paper is as follows. Section 2 provides a critical survey of some of the main contributions proposed in the optimal control theory to compute the economic depreciation of natural resources analytically and thus arrive at a correct welfare measure. These computations are based on shadow or true scarcity prices. However, observed prices generally differ from scarcity prices because of distortions in the economy that make it diverge from the optimal path. Therefore, Section 3 examines the issue of the accounting prices that can be used to determine the economic depreciation of natural capital. Section 4 deals with the implementation of the adjustments suggested by the theory in the national accounts. In particular Section 4.1 examines the treatment of environmental defensive expenditures since different theoretical models take different approaches to this issue. Section 4.2 deals with the estimation of marginal environmental damages and costs. We consider two different methodologies that could be applied to obtain a rough, first approximation of environmental damages and costs at aggregate level, taking into account that the microeconomic evaluation methodologies are too difficult to implement at national level. Section 4.3 reviews the methodologies proposed to calculate natural resource depreciation. Section 5 contains some concluding remarks.

The adjusted NNP as a true welfare measure: the theoretical debate in the literature

The Weitzman model: NNP as welfare measure and the interpretation of the current-value Hamiltonian

Surprisingly enough, most of the debate on environmental accounting started with a seminal paper by Weitzman (1976) which contained no reference to environmental issues. In his paper Weitzman asked why NNP, defined as consumption plus net investment, could be regarded as a good measure of welfare, as commonly accepted by many authors before him. In fact, according to Samuelson's definition of welfare as present value of consumption, consumption and not capital formation is the ultimate aim of economic activity. Why then include investments in the measure of welfare, as NNP does?

As Weitzman himself says (1976, p.159), "if all investments were convertible into consumption at the given price-transformation rates" the maximum consumption which could be maintained for ever without running down the capital stock would be just the NNP, as conventionally measured:

(1) $NNP = C^* + p(dK^*/dt).$[2]

However, it is not possible to convert in reality all investments into consumption. Therefore, $C^* + p(dK^*/dt)$ is not feasible and - a fortiori - not permanently maintainable. To show why this is the case, Weitzman used a very simple diagram (Figure 1).

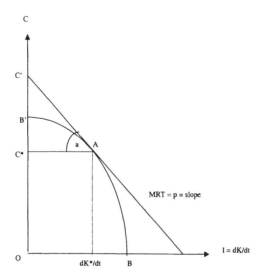

Figure 1: The Weitzman Model

Source: Weitzman, 1976. Legend: MRT = marginal rate of transformation

The distance OC' is the geometric equivalent of real NNP. In fact it is:

(2) $OC' = OC^* + C^*C'$

$p = tg(a) = C'C^*/AC^*$

where tg(a) is the tangent of the angle a (Figure 1). Therefore, from (3):

$C'C^* = p(AC^*) = p(dK^*/dt)$

which, substituted back into (2), yields:

$OC' = C^* + p(dK^*/dt) = NNP.$

Since the economy can only reach the points along the production possibility frontier BB', OB' is "the largest permanently maintainable level of consumption that can *actually* be obtained", while OC' is a "*strictly hypothetical* consumption level at the present time" (Weitzman, 1976 p. 159).

---

[2]   The stars indicate the optimal values of the variables in question.

At first sight, the fact that NNP corresponds to an unfeasible consumption level seems to imply that it cannot be taken as a measure of welfare. However, from a deeper insight into the nature of NNP, Weitzman shows that this notion can be viewed as a welfare measure, although it is a hypothetical consumption level and not a real trajectory of an economic system. In fact, it can be proved that:

$$(4) \quad \int_t^\infty C*(s)\, e^{-r(s-t)}\, ds = \int_t^\infty \left[C*(t) + p(t)\frac{dK*}{dt}(t) \right] e^{-r(s-t)}\, ds$$

Equation (4) states that the present value of consumption along an optimal path (the term on the left-hand side) is equal to the present value of NNP, if maintained constant from time t onwards (the term on the right-hand side). In other words, as Weitzman claims (1976, p.160) "the maximum welfare actually attainable from time t on along a competitive trajectory is exactly the same as would be obtained from the hypothetical constant consumption level" given by NNP. The NNP is thus a "proxy for the present discounted value of future consumption" (Weitzman 1976, p.156) and the "*stationary equivalent of future consumption*" (Weitzman 1976, p.160).

Beside this notion of NNP, Weitzman's theoretical framework provides another possible interpretation of the concept: "NNP is what a social planner would choose to maximise" (Atkinson et al., 1999 p.34). This can immediately be verified by examining the structure of the optimisation problem. As mentioned above, in accordance with Samuelson's definition of welfare, Weitzman takes the present value of consumption as the objective function to be maximised. This is equivalent to assuming a utilitarian framework, with linear utility function:

$(5)\ U(C(t)) = C(t).$

Therefore, a hypothetical social planner would have to solve the following optimisation problem:

$$(6) \quad \underset{C(t)}{Max} \int_t^\infty C(t)\, e^{-rt}\, dt$$

subject to:

$$(7) \quad (C(t), \frac{dK}{dt}(t)) \in S(K(t))$$

$$(8) \quad K(0) = K_0$$

where S(K(t)) is the production possibility set at time t represented in Figure 1 and K_0 the stock of initial capital available at time 0.

Equation (7), which represents a condition of efficiency in production,can also be written as an explicit function:

$$(9) \qquad \dot{K}(t) = \frac{dK}{dt}(t) = f(C(t))$$

Along the optimal path, the current value Hamiltonian $H_c(t)$ corresponding to the above maximisation problem is therefore:

$$(10) \qquad H_c(t) = C*(t) + p(t)f(C*(t)) = C*(t) + p(t)\dot{K}*(t)$$

where the costate variable p(t) is the shadow value of capital.

Equation (10) defines an index that is linear both in consumption and in the investment level. Since the price of consumption is taken as numeraire in the model (i.e. it is equal to one), this linear index is equal to the NNP as measured along an optimal competitive trajectory. Hence, in the simple maximisation problem set forth by Weitzman, *the current value Hamiltonian is the NNP*.

Extending Weitzman's framework to the environmental issue: the Hartwick model

As Pemberton and Ulph (1998, p.1) have pointed out, Weitzman's conclusions hold "in the context of the particular model considered in the paper", as they heavily hinge on specific assumptions. In this regard we can identify three main hypotheses which distinguish Weitzman's model:[3]

1. a *linear utility function*
2. a *fixed interest rate* on the consumption good
3. an explicit *definition of income* as the maximum feasible NNP, namely:

$$(11) \qquad Y(K, p) = \max_{(C,I) \in S(K)} [C + pI]$$

Despite the specific features of Weitzman's framework, many authors have tried to extend his analysis to the environmental context to determine how the NNP should be adjusted to incorporate the depreciation of natural

---

[3] Another assumption which plays an important role is the fact the economy is supposed to be closed. Due to space constraints, we will not examine the case of an open economy in the present paper.

resource stocks. One of the main contributions in this sense is the Hartwick model (1990).

Hartwick examines the optimal growth of an economy with natural resources, which he divides into three main categories: exhaustible resources, renewable resources and environmental capital. Following Weitzman's approach, Hartwick also assumes a utilitarian objective function, but he does not make any specific assumption about the linearity of the function. This raises the following question: if the utility function is non-linear in consumption (C) how can the current value Hamiltonian be linear in its arguments and thus equal to the NNP? [4]

To answer this question, let us first examine Hartwick's maximisation problem and focus attention on the first of the three categories that he investigates: exhaustible resources.

Exhaustible resources In the case of an economy that relies on an exhaustible resource for production, the optimisation problem is as follows:

$$(12) \quad Max \int_0^\infty U(C)e^{-\rho t} dt$$

$$(13) \quad s.t. \ \dot{K} = F(K,L,R) - C - f(R,S) - g(D,S)$$

$$(14) \quad and \ \dot{S} = D - R$$

where:

ρ = intertemporal discount rate

S = stock of the exhaustible resource

R = flow of the exhaustible resource

D = discoveries of new stock of the resource

K = physical capital

L = labour

f(R,S) = cost of resource extraction

g(D,S) = cost of resource exploration.

Observe that (13) corresponds to equation (9) in Weitzman's optimisation problem, but we now have an additional state variable (S) and thus also the additional constraint (14) that determines how the exhaustible

---

[4]   Recall, from (1), that the NNP is defined as a measure that is linear in its arguments (consumption and investment).

resource changes over time. The current value Hamiltonian of the above problem is:

$$(15) \quad H_c(t) = U(C) + \phi(t)[F(K,L,R) - C - f(R,S) - g(D,S)] + \varphi(t)[D-R]$$

If the utility function is non-linear in consumption, the current value Hamiltonian will also be non-linear and thus the equivalence highlighted by Weitzman between $H_c(t)$ and NNP no longer holds.

However, Hartwick takes a linear approximation of the utility function around the point $C=C_0$:

$$(16) \quad U(C) = U(C_0) + U_C(C - C_0)$$

where $U_C = \dfrac{dU}{dC}\bigg|_{C=C_0}$

Substituting (16) into (15), we get:

$$(17) \quad H_c(t) = U_C C + \phi(t)\underbrace{[F(K,L,R) - C - f(R,S) - g(D,S)]}_{\dot{K}} + \varphi(t)\underbrace{[D-R]}_{\dot{S}}$$

where the term $U(C_0) + U_C C_0$ has been omitted since the solution to an optimisation problem is invariant with respect to a constant.[5]

Notice that the marginal utility level computed at the expansion point $C=C_0$ (what he calls U_C) is obviously constant. This has two important consequences.

In the first place, the current value Hamiltonian is linear in C, I and $\dot{S}$ and is equal to the aggregate value of all quantities in the economy (the flow of consumption and the stocks of man-made and natural capital), each valued at its shadow price (1, $\phi(t)$ and $\varphi(t)$ respectively). Therefore, Hartwick takes a linear approximation of the utility function to make his framework analogous to that of Weitzman and thus extend the equivalence of current value Hamiltonian and NNP to his own model. In the second

---

[5]   The linearisation of the utility function that we present here is slightly different from the method followed by Hartwick. This is because we find Hartwick somewhat unclear on this specific point. In fact, after defining the current value Hamiltonian (equation 15) and the corresponding first-order conditions, Hartwick (1990, p.293) claims: "Let us use a linear approximation $U(C)=U_C C$". In fact, this is not a linear approximation of the utility function unless we take the Maclaurin expansion (around C=0), rather than the Taylor expansion (around $C=C_0$). However, even if one takes the Taylor expansion (as we do above), the constant terms do not affect the optimisation problem, therefore they can be omitted in equation (17). This is probably what Hartwick means with the statement above when he takes only $U_C C$ into account.

place, the fact that U_C is constant allows Hartwick to justify measuring NNP in monetary terms. In fact, dividing H_C by U_C we get what Hartwick (1990, p.293) calls the "dollar-value" expression for NNP:[6]

$$(18) \quad \frac{H_c(t)}{U_c} = C + \frac{\phi(t)}{U_c}\dot{K} + \frac{\varphi(t)}{U_c}\dot{S}$$

From the first-order conditions we have:

$$(19) \quad Uc = \phi(t)$$

$$(20) \quad \varphi(t) = [F_R - f_R]\phi(t)$$

where F_R is the marginal productivity of the resource (equal to the resource price in equilibrium) and f_R is its marginal cost of extraction.

Substituting (19) and (20) into (18), it yields:

$$(21) \quad \frac{H_c(t)}{U_c} = C + \dot{K} - [F_R - f_R][R - D]$$

The last addendum on the right-hand side is the correction term that should be measured in green national accounts to quantify the economic depreciation of an exhaustible resource.[7] The term F_R-f_R is obviously the Hotelling rent. To achieve a correct measure of welfare, we should therefore deduct the resource rents given by the product of the reduction in the stock of the exhaustible resource times its shadow price, from conventional measures of NNP (as defined in equation 1).

Pemberton and Ulph (1998) have recently criticised Hartwick's approach for two main reasons. In the first place, they point out that the equivalence of NNP to the Hamiltonian is derived in Weitzman's model from a specific definition of income (assumption (iii) on p.5), whereas Hartwick does not give any explicit definition of income in his paper. In the second place, they argue that the linearisation of the utility function is an "ad hoc" assumption to get a constant marginal utility and thus make the model analogous to that of Weitzman who assumed a linear utility function (assumption (i) on p.5).

To overcome these drawbacks that affect Hartwick's contribution, Pemberton and Ulph assume a non-linear (strictly concave) utility function

---

[6] From the first order conditions, U_c equals $\phi(t)$, that is, the marginal utility of income. Since the current value Hamiltonian is measured in utility terms, dividing $H_c(t)$ by U_c is equivalent to dividing utility by the marginal utility of income, which yields an index measured in income or monetary terms.

[7] In steady state, this term is obviously zero since $\dot{S}=0$.

and solve the optimisation problem without taking any linear approximation of the function. Moreover, unlike Hartwick, they give an explicit definition of income, taking the Hicksian notion as starting point, and try to derive an adjusted measure of NNP that is consistent with that notion in the case of exhaustible resources. In this way Pemberton and Ulph derive the corrected measure of NNP in a more general framework and base it on a notion of income that is well-founded in the economic theory. However, the adjusted NNP obtained by Hartwick is still valid. In fact, Pemberton and Ulph (1998, p.7) show that along the optimum path it is:

$$(22) \quad Y^*_t = C^*_t + \frac{dK_t^*}{dt} - x^*_t R^*_t$$

where Y_t denotes the income level, R_t the resource rent and x_t the flow rate of utilisation of the exhaustible resource at time t.[8] It is easy to verify that the above expression corresponds to equation (21), what they call the National Income Rule suggested by Hartwick. Therefore, the authors conclude that although Hartwick used a *flawed method of correction* of NNP, the *correction term* he obtained is still *valid*.

Hartwick's model can obviously be extended to other cases of exhaustible resources by changing the initial assumptions slightly. Hamilton (1994), for instance, examines the case of a homogeneous exhaustible resource, having N heterogeneous resource deposits with different extraction costs f_i (e.g. oil). As it can be easily verified, in this case we get:

$$(23) \quad NNP = C + \dot{K} - \sum_{i=1}^{N} (F_R - f_{Ri}) R_i$$

where f_{Ri} is the marginal extraction cost in deposit i and R_i is the amount extracted in the same deposit.

Thus, if we have N different deposits, the correction term to be subtracted is the sum of the rents on each resource deposit.

A second modification of Hartwick's analysis can be obtained by changing the discovery cost function g. Hamilton (1994) shows that if g depends on the cumulative discoveries of the exhaustible resource rather than on the remaining stock S, a different correction term emerges from the model. In fact, it is:

---

[8]   Note that x^*_t and R^*_t in equation (22) correspond to $[F_R - f_R]$ and $[R-D]$ respectively in equation (21).

(24) $NNP = C + \dot{K} - (F_R - f_R)R + g_D D$

If we compare this expression with equation (21), we see that the new discoveries are now valued at their marginal cost (g_D) rather than at the Hotelling rental rate ($F_R - f_R$).

Renewable resources So far we have considered the case of an exhaustible resource. But how should NNP be adjusted to account for the depreciation of a renewable resource?

Let us call Z the stock and E the level of exploitation of such a resource, say fish. In this case, Hartwick (1990) argues that the flow E should be treated as a source of utility as it may be consumed directly by the representative agent. Therefore, he sets out the following central planning problem:

(25) $Max \int e^{-\alpha} U(C,E)$

(26) $s.t. \dot{K} = F(K,L) - C - f(E,Z)$

(27) $and \ \dot{Z} = r(Z) - E$

where $f(E,Z)$ is the extraction cost of the renewable resource and $r(Z)$ is its rate of growth. The corresponding current-value Hamiltonian is:

(28) $H_c(t) = U(C,E) + \phi(t)[F(K,L) - C - f(E,Z)] + \varphi(t)[r(Z) - E]$

Following the procedure described above for an economy with an exhaustible resource, Hartwick (1990) replaces the utility function with its linear approximation[9] and then divides $H_c(t)$ by the constant term U_c to express the NNP in dollar value terms. This yields:

(29) $\dfrac{H_c(t)}{U_c} = C + \dfrac{U_E}{U_c} E + \dfrac{\phi(t)}{U_c} \dot{K} + \dfrac{\varphi(t)}{U_c} \dot{Z}$

From the first-order conditions:

$U_c = \phi(t)$

$\varphi(t) = U_c[\dfrac{U_E}{U_c} - f_E]$

substituting into (29) we obtain the accounting rule for the case of renewable resources, that is:

---

[9] The considerations pointed out in footnote 5 also apply in the present case, the only difference being that utility is now a function of two variables.

25

$$(30) \quad \frac{H_c(t)}{U_c} = C + \dot{K} + \frac{U_E}{U_c}E + [\frac{U_E}{Uc} - f_{\varepsilon}]\dot{Z}$$

where $\dfrac{U_E}{U_C}$ is the market price of the renewable resource and f_E its

marginal extraction cost (e.g. the cost of fishing one additional fish). As in the case of any other capital good, the adjustment of NNP to account for renewable resources is therefore in rental form (i.e. price minus marginal cost). More precisely, traditional NNP ($C + \dot{K}$) should be corrected by adding the value of the flows of the renewable resource and the change in its stock valued at the rental rate.[10]

Environmental capital goods and pollution Let us now turn to the third category of natural resources examined by Hartwick: the environmental capital goods, such as airsheds and watersheds, the value of which is reduced by pollution. Hartwick assumes that there exists an *indirect* mechanism of pollution control via production: higher production levels imply higher pollution that in turn reduces the output Y for a given level of labour and capital. The production function is then:

$$(31) \quad Y = F(K, L, X)$$

where X is the pollution stock and dF/dX < 0.

Nature tends to absorb part of the pollution stock X in the atmosphere, b being the natural rate of absorption. Given these assumptions, the equation of motion of X can be written as follows:

$$(32) \quad \dot{X} = \gamma F(K, L, X) - bX$$

where γ is the ratio at which production increases pollution.

The social planner is now confronted with the following dynamic maximisation problem:

---

[10] As Hartwick points out (1996, p.295), the change $\dot{Z}$ in the resource stock is probably negative in a world where both population and income grow substantially over time. Hence, the last term in equation (30) measures the economic depreciation of the renewable resource and it should be netted-out from NNP to get a true measure of national income.

(33) $$Max \int_0^\infty e^{-\rho t} U(C) dt$$

(34) $$s.t. \ \dot{K} = F(K, L, X) - C$$

(35) $$and \ \dot{X} = \gamma F(K, L, X) - bX$$

Replacing the utility function by its first order approximation and the costate variables by the corresponding first order conditions, Hartwick ends up with the following corrected measure of NNP in monetary terms:

(36) $$\frac{H_c(t)}{U_c} = C + \dot{K} + \left[\frac{\rho - F_K}{\gamma F_K} \right] \dot{X}$$

where F_K is the marginal productivity and γF_K is the marginal pollution of one additional unit of capital.

As Hartwick points out, the reciprocal of γF_K is the increase in man-made capital foregone because of pollution, which reduces the economy's productivity. Thus, in the present case, the depreciation term is the increase in the stock of pollution ($\dot{X} > 0$) multiplied by the negative effect of this increase on the investments in physical capital ($1/\gamma F_K$).

Besides this indirect control mechanism "via the output decision of the producers" (Hartwick 1990, p.298), pollution can also be controlled *directly* via abatement expenditures. The cost f of these expenditures, which depends on nature's capacity b to absorb pollution, reduces the amount of output left for investments:

(37) $$\dot{K} = F(K, L, X) - C - f(b)$$

Replacing equation (34) with (37) in the optimisation problem and solving as described above, Hartwick gets:

(38) $$\frac{H_c(t)}{U_c} = C + \dot{K} - \frac{f'(b)}{X} \dot{X}$$

The higher the natural rate of absorption b (that is, the faster pollution evaporates by natural regeneration of environmental capital), the higher the marginal cost of increasing b by investing in abatement capital. Hence:

$f'(b) > 0$ which implies $-\dfrac{f'(b)}{X} \dot{X} < 0$.

If pollution is increasing over time ($\dot{X} > 0$), we should therefore deduct from conventional NNP the cost of reducing pollution by $\dot{X}$ or equivalently the investment in physical capital foregone ($f'(b)$) to invest in abatement capital that reduces pollution by $\dot{X}$.

Finally, Hartwick makes a further modification to the original optimisation problem (33)-(35), by assuming that the agent's utility is a function not only of consumption, but also of the flow of pollution $\dot{X}$:

(39) $U = U(C, \dot{X})$ where $U_c > 0$.

Replacing equation (33) with (39) and (34) with (37), the analytical problem becomes:

$$Max \int_0^\infty e^{-\rho t} U(C, \dot{X}) dt$$

$$s.t. \quad \dot{K} = F(K, L, X) - C - f(b)$$

$$and \quad \dot{X} = \gamma F(K, L, X) - bX$$

Taking the linear approximation of the objective function and substituting from the first-order conditions into the current-value Hamiltonian, Hartwick derives the following expression for the NNP:[11]

$$(40) \quad \frac{H_c(t)}{U_c} = C + \dot{K} - \left[\frac{U_{\dot{x}}}{U_c} + \frac{f'(b)}{X} \right] \dot{X}$$

Hence, if pollution adversely affects both production and utility, the appropriate correction term is the difference between the price of extra pollution ($\frac{U_{\dot{x}}}{U_c}$) and its marginal cost ($\frac{f'(b)}{X}$). Hence, the usual accounting rule applies also in this case: to obtain a correct measure of NNP, deduct the economic depreciation of the environment in its rental form.

A more comprehensive approach: the Dasgupta-Kriström-Mäler model and differences in the correction term

As we have seen, Hartwick dealt with the three kinds of natural resource separately and showed that the correction terms that emerge from independent models share the same accounting principle. However, some authors prefer to develop a single model that combines all kinds of natural resources simultaneously to get a corrected measure of NNP. Among them,

---

[11] Note that it follows from the first order conditions that the term in brackets is positive.

a major contribution is that of Dasgupta, Kriström and Mäler, (henceforth DKM, 1997). As we show below, the accounting prescriptions that emerge from DKM differ from those achieved by Hartwick. This can be explained by the different analytical settings of these two papers. Different models obviously lead to different magnitudes to be netted out. In fact, as we pointed out above, the formula of the correction term depends on the expression of the current-value Hamiltonian, which in turn is determined by the features of the optimisation problem, namely, the arguments of the utility function and the constraints to the maximisation.[12]

The first major difference between Hartwick and DKM lies in the objective function. In DKM the agent's utility is not only a function of consumption C of the good produced by the economy, but also of leisure l and the environment. The latter enters the utility function in three different forms:

- the consumption E of a renewable resource (e.g. fuelwood from forests which can be used to cook)
- the stock K_2 of an environmental capital good (e.g. clean air)
- the flow A of environmental amenities.[13]

The objective function in DKM is thus:

$$(41) \int_0^\infty e^{-\rho t} U(C, l, E, K_2, A) dt$$

where utility increases in all its arguments. Call O the vector of the arguments of the utility function: $O = [C,l,E,K_2,A]$

As mentioned above, a second difference between Hartwick and DKM is that the latter deal with all forms of natural capital simultaneously. We thus have four different kinds of capital stock in the model:

---

[12] As Hartwick (1990, p. 303) states: "Imbedded in this (current-value) Hamiltonian were the formulas for netting-out the 'consumption' of natural resource stocks over the accounting period".

[13] Note that only the first form (consumption of renewable resources) is present in the utility function assumed by Hartwick. Moreover, the fact that leisure is among the arguments of the objective function implies that labour is now a choice variable in the model.

1. the stock of man-made capital K_1
2. that of environmental capital K_2
3. the stock of the renewable resource K_3
4. that of defensive capital K_4

A different kind of labour Lj (j = 1…4) is associated with each form of capital.

Pollution P reduces the environmental amenities A and the stock of environmental capital K_2 (e.g. clean air or water). However, society can intervene to counter the effects of pollution in two ways:

It can make defensive expenditures R to restore environmental amenities damaged by pollution. It can invest in defensive capital K_4 that reduces the level of pollution.

We then see that pollution increases with production, but decreases with defensive capital:

(42) $P = P(Y, K_4)$ where $P_Y > 0$, $P_{K4} < 0$ and $Y = F(K_1, L_1)$

whereas the environmental amenities A decrease with pollution and increase with defensive expenditures:

(43) $A = A(P, R)$ where $A_P < 0$, $A_R > 0$

Substituting (42) into (43), we get:

(44) $A = A[P(F(K_1, L_1), K_4) ; R]$

Notice that, unlike Hartwick, DKM introduce a twofold defensive intervention (R and K_4) in their model. This intervention affects the agents' utility in two ways. First, R and K_4 influence the level of environmental amenities A via equation (44), which in turn affects the agents' well-being via equation (41). Second, expenditures in R and K_4 reduce consumption (and thus the utility level) since:

$$C = F(K_1, L_1) - \dot{K}_1 - \dot{K}_4 - R$$

The optimal trajectory of the economy in DKM is determined by maximising (41) subject to (44) and the equations of motion of the capital stocks. In this case, the corresponding current-value Hamiltonian is:

(45) $H_C = U(C, l, E, K_2, A) + p\dot{K}_1 + q\dot{K}_2 + r\dot{K}_3 + s\dot{K}_4 + vA(P(F(K_1, L_1), K_4), R)$

where p, q, r, s and v are the costate variables associated with the constraints.

As the authors show, the above Hamiltonian satisfies the following important condition:

$$(46) \quad \int_t^\infty e^{-\rho(s-t)} U*(\bullet)ds = \int_t^\infty e^{-\rho(s-t)} H_c * ds$$

$$where \quad U*(\bullet) = U(C^*, L^*, E^*, K_2^*, A^*)$$

The above equality states that the current-value Hamiltonian expresses a utility level which, if maintained constant from t to infinity, would have the same present discounted value of utility along an optimal path from t onwards. As Heal (1998) has underlined, this result is very general and does not depend on the model we are analysing.[14] This is confirmed by comparing equations (46) and (4): the former is just a reformulation of the latter with the optimal value of Weitzman's linear utility function (C*) on the left-hand side replaced by that of a general utility function (U*). We can therefore reformulate Weitzman's conclusion and claim that the NNP is *the stationary equivalent of the future utility stream*.[15] Hence, as Heal (1998, p.167) states, "the (current-value) Hamiltonian is a measure of the *sustainable utility level* associated with an optimal path" and it is therefore a measure of welfare.

However, Heal (1998, p. 167) also points out that "the absolute value of the Hamiltonian has no significance". In fact, when implementing a specific policy, what matters is the effect that the policy has on the sustainable utility level (and thus on social welfare) rather than the level itself. Therefore, what we should examine is "whether the policy increases the value of a linear approximation to the Hamiltonian" (Heal 1998, p.167).

This is exactly what DKM do. In order to evaluate the effect of a small perturbation (i.e. a project that leaves prices unchanged), they take the first order Taylor approximation of the current-value Hamiltonian around the vector of optimal values O*:

$$(47) \quad LH_c = H_c(O^*) + \left.\frac{dH_c(O)}{dO}\right|_{O=O^*} (O - O^*)$$

where LH_c is the linearised current-value Hamiltonian.

---

[14] In fact, Heal (1998) shows that this equality holds true in any problem of optimal dynamic use of natural resources.

[15] Recall that in Weitzman the NNP is "the stationary equivalent of future consumption" (see page 4).

Omitting the constant (which does not affect the outcome of the maximisation), DKM take the term $\left[O \dfrac{dH_c(O)}{dO} \bigg|_{O=O^*} \right]$ as the correct measure of NNP. After a few mathematical manipulations, this term can be written as:

$$(48) \quad NNP = U_C C + p\dot{K}_1 + \underbrace{r\dot{K}_3}_{1} + \underbrace{U_X X}_{2} + \underbrace{q\dot{K}_2}_{3} + \underbrace{U_L \left(\sum_{i=1}^{3} L_i \right)}_{4} + \underbrace{s\dot{K}_4}_{5} + \underbrace{U_A A}_{6}$$

where X is the flow of renewable resources (e.g. the output of fuelwood).[16]

This result raises two important considerations. In the first place, as pointed out above, DKM (1997) as well as Heal (1998) call NNP *the linearisation of the whole current-value Hamiltonian*. This contrasts with the approach in most of the literature which, following Hartwick (1990), defines NNP as *the current-value Hamiltonian obtained by replacing only the utility function with its linear approximation*.[17] Therefore, all authors resort to some kind of linearisation, since NNP is a linear index. However, different authors linearise different functions and this might lead to diverse conclusions as to the term to be netted-out. This implies that all authors agree on the fact that the current-value Hamiltonian is a measure of sustainable utility (as suggested by equation (46)), but not on how to compute the NNP. In the second place, the correction term suggested by DKM goes beyond that of Hartwick. According to equation (48), the correct measure of NNP is the traditional NNP (consumption plus investments) plus

1. the appreciation (minus the depreciation) of renewable resources ($r\dot{K}_3$)
2. the value of the flow of renewable resources as measured at its shadow price ($U_X X$)

---

[16] As in Hartwick (1990), to obtain the NNP in monetary terms it is sufficient to divide the right-hand side of equation (48) by the marginal utility of consumption U_c.

[17] Had we linearised only the utility function rather than the whole Hamiltonian, the NNP in DKM would be (omitting the constants):

$$NNP = U_C C - U_l \left(\sum_{i=1}^{4} L_i \right) + U_E E + U_A A + U_{K_2} K_2 + p\dot{K}_1 + q\dot{K}_2 + r\dot{K}_3 + s\dot{K}_4 + vA\left[P(F(K_1, L_1), K_4), R \right]$$

3. the appreciation (minus the depreciation) of environmental capital $(q\dot{K_2})$

4. the shadow wage bill $((U_{Li})\sum_{i=1}^{3}L_i)$

5. the value of changes in defensive capital ($s\dot{K_4}$)

6. the value of changes in the flow of environmental amenities ($U_A A$)

Like in Hartwick, every component of the corrected NNP is valued at its shadow price. Moreover, there is a one-to-one relationship between some terms in (48) and those determined by Hartwick. In fact, terms a) + b) above correspond to the correction measure $\frac{U_E}{U_c}E + [\frac{U_E}{Uc} - f_E]\dot{Z}$ in equation (30) computed by Hartwick for the case of renewable resources. Similarly, term c) above corresponds to $\left[\frac{\rho - F_K}{\gamma F_K}\right]\dot{X}$ in equation (36), that is, Hartwick's correction factor for the case of environmental capital.[18]

However, equation (48) embraces some additional terms (d, e and f) with respect to the correct measure of NNP derived by Hartwick.

The presence of term (d) in equation (48) indicates that labour income should not be included in a welfare measure. Mäler (1995) provides an intuitive explanation for this unexpected result. Since agents can choose their optimal allocation of time between labour and leisure, in equilibrium they are indifferent to one more hour of leisure or work. Thus in equilibrium, "the gains from increased production of consumer goods are completely offset by the costs from reductions in leisure" (Mäler 1995, p.140). This result hinges on the obviously unrealistic assumption of perfect-clearing labour markets.

However, Mäler argues that the same holds true even if we relax this assumption, provided we take the agent's marginal reservation wage as the accounting price for labour. In fact, if the reservation wage is zero, the shadow wage bill will also be zero, which again leads to the exclusion of labour income from the national product.[19]

---

[18] Note that K_2 in equation (48) and X in (36) can be thought of as being opposites: K_2 represents the stock of clean air, whereas X is the stock of pollution.

[19] Assuming a zero reservation wage implies that the worker is willing to accept a job as long as she receives any positive wage. This again may be a rather strong assumption. For instance, an agent will probably *not* accept a job if the wage is below the cost of going to work or paying a baby-sitter to look after the children. Therefore, if our interpretation of Mäler's argument is correct, one might still want to include labour income in a measure of

According to DKM the investments in the stock of defensive capital (term 5 in equation (48)) should also be included in the national income to arrive at a correct measure of welfare. It is worth pointing out, however, that expenditures R to restore environmental amenities damaged by pollution are not in (48). Therefore, as Mäler (1991, p.6) argues, current defensive expenditures should not be deducted from the NNP to get a "true" welfare measure.[20]

Finally, the flow of environmental amenities (term f in equation (48)) should also be evaluated in the NNP. This is an additional term with respect to the NNP computed by Hartwick, who did not include environmental beauty in the agents' utility function. However, natural resources are not only an input in the production function, but also a source of direct well-being (think, for instance, of the pleasure that derives from observing a beautiful natural landscape). Therefore, any loss in environmental amenities should imply a reduction in the welfare measure.

So far we have considered how GNP should be adjusted to take economic depreciation of natural resources into account. In all cases, the adjustment term is computed using shadow values for all changes in capital stock, assuming that the economy moves along an optimal path. However, real economies are characterised by failures in property rights that make them diverge from the production possibility frontier. How can we determine the correction term when observed prices differ from true scarcity prices? Can we still rely on the current-value Hamiltonian as a useful benchmark to determine the adjustment term and thus the corrected NNP?

We will address these questions in the next section where we examine the difficulties that arise in practice when one tries to implement the theoretical arguments seen above to compute a true welfare measure.

The problem of accounting prices

The linearised version of the current-value Hamiltonian is, in fact, a linear combination of values, i.e. prices times quantities, like the national accounting NNP, but unlike the latter, it may be a measure of social welfare. If we could modify the existing accounting version of NNP to take the

---

national product. However, we will not discuss here this interesting argument proposed by Mäler, as it is beyond the scope of the present paper.

[20] We will return to this point in the next section, where the issue of defensive expenditures will be examined more closely.

adjustments suggested by the theory into account we could have a measure of social welfare and we could compare levels of welfare in different periods of time.

The problem with the implementation of the ideal welfare index is that the national accountant uses observed prices and quantities that are generally distorted or non-scarcity prices, whereas the ideal index of welfare is based on accounting prices. These can be estimated in a number of ways. One way is to use prices that sustain an optimal plan, i.e. shadow or scarcity prices. A second is to use "local" prices. DKM (p.130) explain the problem with some simple diagrams.

Suppose we have an economy consisting of two consumer goods and one individual. Let us assume that X and Y in figure 2 are the consumption goods and TT' is the production possibility frontier. The individual's well-being is given by the utility function $U(X, Y)$ and II' is the individual's indifference curve which is tangent to TT'. The slope of the common tangent at point A defines the optimal prices p_x, p_y. Thus, at any production point we can define NNP= $p_xX + p_yY$. This is NNP computed using optimal prices. Let us now assume that the economy is at B (a point on the production frontier). We want whether or not a move from B to C is an improvement in the individual's well-being. Since the bundle C is on a higher indifference curve, C is preferred to B and the utility (welfare) associated with C must be higher than the utility associated with B. Thus, the use of optimal prices to evaluate bundle B and C results in the NNP (as a measure of welfare) at C being higher than the NNP at B.

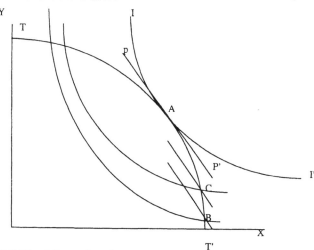

Figure 2: NNP with optimal prices
Source: Dasgupta-Mäler, 1997

However this result depends heavily on the assumption that a shift from B to C is small and along the frontier. Suppose that the project actually causes a large move along the frontier, as in the figure below. In this case the NNP increases from B to C (because we are on a higher budget line for the same prices), but the agent's utility decreases. The same considerations apply if we use local prices to evaluate a big move along the production possibility frontier.

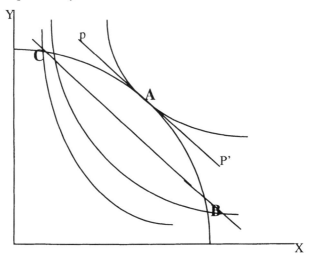

Figure 2a: Higher NNP but lower utility level
Source: Dasgupta-Mäler, 1997

Let us now suppose that we are not on the production possibility frontier, but on a point in the production possibility set (like point B in Figure 3) as it often occurs in real economies. In this case, optimal prices are inappropriate, because the tangent condition is not satisfied. We can use local prices, i.e. the individual marginal rate of substitution at that point. Suppose that (X, Y) is the bundle chosen by the individual. Suppose that there is a small change in consumption or production: ΔX and ΔY in discrete terms. Since this is a small project, relative prices do not change and the resulting change in the individual's well-being is $U_X \Delta X + U_Y \Delta Y$, where $U_x = MU_x$ and $U_Y = MU_Y$.[21]

---

[21] MU_x being the marginal utility of X in discrete terms: $\Delta U / \Delta X$.

This change involves an increase in well-being for ($U_X\Delta X$ + $U_y\Delta Y$)>0 and a decrease for ($U_X\Delta X$ + $U_y\Delta Y$)<0. Thus, the marginal evaluations of the individual can be used as accounting prices. In other words, NNP evaluated on the basis of current marginal valuation is an appropriate measure of social well-being.

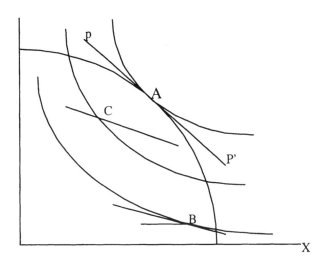

Figure 3: NNP with local prices
Source: Mäler, 1995

In Figure 3, the individual's marginal rate of substitution at B: -(MU_x/MU_Y) is used as "local" prices to evaluate NNP at C. Since C lies on a higher indifference curve, $NNP_C>NNP_B$.[22]

Unfortunately the national accountant does not find himself in A position to use optimal prices or "local prices". He uses current prices, i.e. observed prices, that may be distorted or non scarcity prices. This means that if p_x and p_y are current observed prices, they may be different at B and C, so that we do not get an appropriate measure of the variation in social well-being.

---

[22] As is the case of optimal prices, this conclusion depends upon the form of the indifference curve.

The dilemma is that the ideal NNP is very difficult to calculate and true economic welfare can only be associated with these ideal entities, whereas NNP based on observed prices is relatively straightforward to calculate, but cannot claim to be the basis of a welfare measure.

What we can do is to use the observed prices and quantities, where they exist, assuming that the optimal or local prices are not too remote from the observed prices. With them we try to estimate optimal or local prices for those variables entering the ideal measure of social well-being that do not have an observed market price, such as environmental damages or benefits. This procedure would not however add much rigour to our measurement exercise when all other adjustments are estimated at current prices.

Accounting for theoretical adjustments to NNP

The accounting treatment of defensive expenditures

As derived in section 2 NNP does not include adjustments for defensive environmental expenditures implying that, if they are part of NNP, they should not be subtracted. However it is often claimed that defensive environmental expenditures should be deducted from an adjusted measure of NNP, because they do not represent an improvement in human well-being. What then, is the correct approach?

Defensive expenditures are defined as those expenditures borne by households, firms and the public sector that are made in order to "defend" households from environmental degradation (Mäler, 1990, p. 49). The concept includes both "preventive expenditures" and "replacement costs". Hueting (1980) and Olson (1977) have defined defensive expenditures as those expenditures that are related to a negative externality that follows production or consumption activities and among those, environmental defensive expenditures are related to environmental externalities, i.e. to loss of environmental quality. This last class of defensive expenditures can be classified into four categories (U.N. 1993 p. 42):

a) *Environmental Preventive Measures:* changes in the characteristics of goods and services; changes in consumption patterns; changes in production techniques; treatment or disposal of residuals in separate environmental protection activities; recycling; prevention of degradation of landscape and ecosystems;

b) *Environmental Restoration:* reduction or neutralisation of residuals; changes in spatial distribution of residuals; support of environmental assimilation; restoration of ecosystems, landscape and so on;
c) *Avoidance of damages and of environmental deterioration:* evasion activities; screening activities;
d) *Treatment of damages caused by environmental repercussions:* repairs of buildings, production facilities, historical monuments and so on; additional cleaning activities; additional health services; other compensatory activities.

The treatment of defensive expenditures varies according to the economic subject who bears them - we can distinguish between firms' environmental expenditures and environmental expenditures borne by households or the public sector – and, as far as firms' environmental expenditures are concerned, to the fact that they may be "current" defensive expenditures or "defensive capital", i.e. man-made capital that has an environmental defensive function.

It has often been argued that defensive expenditures should be deducted from national accounts, as they do not correspond to an improvement in human well-being. More specifically, those borne by firms are intermediate in nature and are therefore not part of the Value Added (VA) or NNP. Mäler (1990) shows that this argument is not correct because defensive expenditures are included in the VA of selling companies and thus contribute to NNP like any other intermediate good. As to defensive expenditures by households, the counter argument against their deduction is that households buy goods and services in order to achieve an improvement and, in this respect, there is nothing that makes defensive expenditures different from other expenditures. Mäler (1990, p. 49) is quite clear on this point:

> When households spend on extra clean up because of air pollution, this indicates a strong demand on cleaner air so that the value of the improved air means an improvement in human well being. The conventional accounts, even in this case, would show no change in NNP, but if defensive expenditures would be deducted from final demand, NNP would actually fall: an absurd result.

In theoretical models of welfare accounting this point is formalised in different ways. Sometimes there exists a flow of pollution that is an argument of the production function, as in Hartwick (1990) or in Beltratti (1995) for instance, and the value of "defensive capital" approximates the change in the stock of pollution over time. Otherwise, there may be a flow of environmental defensive expenditures that have an impact on the social welfare function through one of the other determinants of welfare, such as environmental quality or environmental amenities, as in DKM (1997).

Let us again consider two of the theoretical models on welfare accounting and the environment: Hartwick (1990) and DKM (1997) and their treatment of environmental variables.

Hartwick considers the environment from a capital-theoretical perspective and tries to express changes in environmental capital stocks as "economic rents" as it happens for any other capital good. Hartwick considered a stock of environmental pollution (section 2.2.3 equation (31)) as a negative input in the production function of firms:

$$Y = F\left(\underset{+}{K}, \underset{+}{L}, \underset{-}{X}\right)$$

We can think of X as being negatively correlated with a stock of environmental capital $S_1 = S_0 - X$ where S_0 is the initial stock of environmental capital. Thus we can write Hartwick's production function as $Y = \left(\underset{+}{K}, \underset{+}{L}, \underset{+}{S_1}\right)$. As we saw from equation (32), the flow of pollution is expressed as the change in time of the pollution stock, which depends on the level of production:[23]

$$\dot{X} = \gamma F\left(K, L, X\right)$$

Thus production is decreasing in the pollution stock and the flow of pollution over time is increasing in the level of production.[24] In Hartwick the value of the change in the pollution stock represents depreciation of environmental capital evaluated in dollar terms as explained in section 2.2.3. $\dot{X} > 0$ implies degradation of environmental capital. Since pollution can only be controlled indirectly through the production decision (greater production implies an increase in X and thus a decrease in Y), the depreciation of environmental capital is expressed as missed investment in reproducible capital K due to decreased production. When firms invest in abatement technologies, they neglect investment in reproducible capital K in favour of pollution abatement capital, the value

---

[23] We ignore the rate of natural regeneration of the environmental stock for the sake of simplicity.

[24] This is reasonable: as Beltratti (1995, p.3) has pointed out: "firms emit in order to produce and suffer themselves from pollution; this happens in very polluted areas such as Mexico City where there is evidence of decreased productivity of workers as a consequence of bad environmental conditions."

of which represents depreciation of environmental capital. Thus, in this model the value of "defensive capital" (defensive expenditures in the capital account) approximates the value of the change in the stock of pollution X. Capital defensive expenditures should therefore be deducted from the measure of aggregate welfare, because they represent the value of depreciation of environmental capital.

As to current defensive expenditures borne by households and the public sector, in the Hartwick model we do not have a flow of current defensive expenditures that enters the social welfare function directly or indirectly. This is because current defensive expenditures are usually introduced in the model as an argument of an "environmental quality" or "environmental amenities" function that is absent from the Hartwick model.

A more general treatment of the environment that also includes current environmental defensive expenditures is found in the DKM model as explained in section 2.3. Here renewable resources are modelled as a stock variable K_3. Moreover there is a stock of environmental capital K_2, such as clean air or water, which does not enter the production function, as can been seen from equation (41). Instead, K_2 enters the aggregate welfare function directly and its variations are considered as depreciation of environmental capital.

There is also a flow of pollution P which increases with the level of production Y, as in Hartwick, and decreases with the stock of defensive capital, K_4, which is here included explicitly in the model, as shown in equation (42). Current environmental defensive expenditures enter the model through a flow of environmental amenities (otherwise called "environmental quality"), which increases with R, the flow of current defensive expenditures, and decreases with P as shown in equation (43). Environmental amenities, A, enter directly the aggregate welfare function in the following way:

$$U\left\{ A\left[P\left(\underset{-}{K_4}, \underset{+}{Y} \right), \underset{+}{R} \right] \right\}$$

In this model we therefore have a stock of defensive capital that has an indirect impact on welfare, as in Hartwick, and a flow of current defensive expenditures, R, which also has an indirect impact on welfare through the flow of environmental amenities.

In both models there is the problem of how to evaluate environmental benefits or damages arising from changes in $\dot{X}$ (in Hartwick) or A (in

DKM) that must be included in the correct welfare measure. The difference between the two models is that DKM introduces current defensive expenditures through the flow of environmental amenities A, whereas Hartwick does not. Since R is included in the final demand (and are thus already part of NNP), there is no need for an additional treatment of current defensive expenditures when the value of changes in A is accounted for. The correct measure of welfare according to the DKM model is thus given by the terms a to f described in section 2.3. When the term, f, the value of changes in the flow of environmental amenities, is computed, current defensive expenditures, that are already included in the final demand, should not be deducted.

From the above analysis we can draw some conclusions about the treatment of defensive expenditures in national accounts.

1. Defensive expenditures by firms are intermediate inputs into the production function and, as such, they are part of the VA of firms that produce them.
2. When capital defensive expenditures borne by firms are used to reduce pollution they can be taken as an approximation of the depreciation of environmental capital. In these cases capital defensive expenditures should be subtracted from a welfare measure.
3. When current households' defensive expenditures are an argument in the flow of environmental amenities or environmental quality (as in DKM), their welfare effects are reflected in the variation in the flow of environmental amenities and there does not seem to be a need for a specific treatment of current defensive expenditures in national accounts.

A somewhat different interpretation of the accounting treatment of households' defensive expenditures is given by Hamilton (1996, p.25). Hamilton shows that, when environmental benefits are priced at the marginal defensive cost, not only defensive expenditures by households should not be deducted, but the welfare measure should include both households' defensive expenditures and the value of environmental benefits to households priced at the marginal defensive cost. This implies that the measure of welfare increases.

Environmental damage

As we have seen in section 2.3 pollution may alter both the flow of environmental amenities and the stock of environmental capital. There is therefore a need to account for changes in utility due to changes in environmental amenities and environmental capital. This raises the problem of how to estimate environmental damage or benefits.

Again one should distinguish between firms and other economic agents.

Indeed, as far as the damage harms companies, it is already included in conventional accounts via variations in production. There is therefore no need to explicitly include current reduction to the production level caused by environmental damage. On the other hand, damage to households which decrease utility and thus welfare are not included in the accounts and it is therefore necessary to obtain data on environmental damage.

There is now a large literature on the measurement of willingness to pay for environmental amenities that includes methods such as the travel cost method, hedonic pricing and contingent valuation. However, as underlined by Hamilton (1996, p. 29):

> Much of the valuation literature is concerned with valuing individual sites or environmental assets. The question of how to sum across the myriad environmental assets within a country in a consistent, non duplicating manner is an unanswered question.

Moreover, as Mäler (1996, p. 13) has noted, these techniques have been developed for use in particular situations where a cost-benefit analysis is needed for making a decision and not for routine production of national product estimates. For this last purpose their cost would probably be prohibitive. Thus, less precise but robust techniques are needed to evaluate environmental damages on an aggregate scale.

One possibility is to use defensive expenditures as an approximation of environmental damage and subtract them from the accounts. There is, however, no foundation in the procedure of approximating environmental damage through the level of defensive expenditures. To the contrary, we can think of examples where defensive expenditures are not proportional to damage.[25]

An alternative, proposed by Mäler (1996), is to treat environmental damage in the same way as production by the public sector is treated in

---

[25] Let us consider of an oil spill in the ocean. The cleaning and restoration cost of the oil spill is easy to quantify and possibly limited in amount, whereas the environmental damage is presumably much greater.

national accounts. Also in this case we have production of goods and services that are not valued in the market. National accountants have solved the problem by looking at the cost of production in the public sector, mainly the cost of labour. Using the same idea to estimate the value of environmental damage requires politically determined targets for environmental quality and the assumption that these targets reflect social preferences as regards environmental quality. In this case, the cost of reaching the politically determined standard for environmental quality would be an approximation of environmental damage. The situation is represented in figure 4 (from Mäler (1996)). Let us assume that MAC is the marginal abatement cost of polluting emissions and MEC is the marginal external cost or marginal damage from emissions. The total abatement and external cost is minimised when the marginal abatement cost equals the marginal damage cost, i.e. at point E. If current emissions are at D, it is desirable to reduce emissions' to C. The total cost of doing so is given by the area CDEA, and the reduction in environmental damage from emission reduction by the area ECDB.

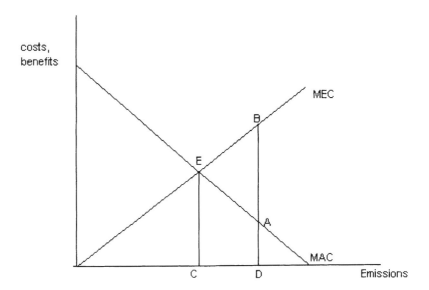

Figure 4: Environmental Damages and Abatement Costs
Source: Mäler, 1996

The two areas are correlated because an increase in total environmental damage due to an increase in emissions is accompanied by an increase in the total abatement cost of achieving the optimum level of emissions. Thus, the cost of achieving the target reduction of emissions is a rough approximation of the environmental damage.

This approach relies on a few assumptions:

a) that politically determined targets for environmental quality exist;
b) that politically determined targets are the only expression of true social preferences we have;
c) that the marginal damage curve is linear.

Oskam (1993) proposed a related approach for the Netherlands and estimated marginal external costs of agricultural production for a number of years. Suppose we are trying to estimate the level of marginal environmental cost[26] of a negative external effect caused by emissions from polluting factors of production (like chemical fertilisers and pesticides). Figure 5 (from Oskam), shows a MEC curve, assumed to be linear and increasing with increasing emissions. Two other curves are drawn: the Marginal Net Private Benefits (MNPB) curve and the MAC curve similar to the one in Figure 4. In an equilibrium situation with no charge for external costs, the producer would choose the level of emissions at which marginal private costs equal marginal private benefits, i.e. MNPB=0, because at this point profit is maximised. The MAC curve may be high or low according to the type of external effect in question. For instance, in agriculture where producers are not even aware of their own level of emissions and abatement plants to reduce leakage of chemicals may be extremely costly to implement, MAC are likely to be very high. To represent the fact that the position of the MAC curve depends on the type of pollution, we draw two MAC curves corresponding to high (MAC_1) and low (MAC_2) marginal abatement costs of emissions.

---

[26] In this section, we assume that marginal environmental and marginal damage costs coincide, although this is not necessarily the case.

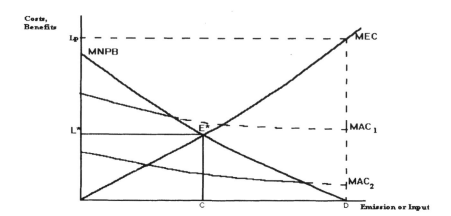

Figure 5: Marginal external costs (MEC), marginal net private benefits (MNPB) and marginal abatement costs (MAC) of an emission

Source: Oskam, 1993

If the most efficient way of reducing emissions is used, the relevant part of MAC is that below the MNPB curve. This is because the producer would not use a quantity of input x for which the abatement cost is higher than the marginal benefit of production. He would rather reduce the amount of input x used. Thus, abatement costs sometimes determine the optimal level of emissions (as assumed, for instance, by Mäler), but under other conditions (when the MAC curve is very high) the MNPB curve is relevant. Unlike the MAC curve, the MNPB curve may be quite easy to estimate, as explained below. When this is so, the optimal level of emissions E* reveals a point on the MNPB and the MEC curves. If we assume that the MEC curve runs through the origin and has a particular form (not necessarily linear), we can generate it.

How can the MNPB and the MEC curves associated with level c of emissions (i.e. L*) be derived? To determine the optimal emission level, c, we can assume consistent decision making by the government in choosing a level of emissions satisfactory from a social point of view. It could be, for instance, the standard level of a polluting input established by the European legislation.

The crucial assumption is, of course, that the chosen level of emissions is the socially optimal one.

The MNPB curve represents marginal profits to the producer and, under perfect competition, also that part of the demand curve of the polluting input x above the input market price p. Let us focus on this part of the MNPB curve, shown in Figure 6.

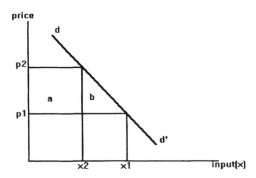

Figure 6: The demand curve for an input
Source: Oskam, 1993

Here the demand function for input x has a linear representation in the form:

$$X = \beta_1 + \beta_2 p$$
$$\beta_1 > 0$$
$$\beta_2 < 0$$

In order to induce the producer to reduce the quantity of x from x_1 to x_2 (from D to C in Figure 5) the price must be increased from p_1 to p_2 (L* in Figure 5). Thus, by knowing the demand curve we can estimate the change in the consumer surplus associated with the increase in input price necessary to achieve the desired reduction. This decrease in the consumer surplus (in this case, the consumer is a producer) is given by area b in Figure 6 (which corresponds to area E*CD in Figure 5) and represents the utility loss due to the reduction in consumption of x.

The information about parameters β_1 and β_2, needed to derive the demand curve, is simple to obtain when we know the average price elasticity of demand for input x ($\bar{e} = \dfrac{dx}{dp} \cdot \dfrac{\bar{p}}{\bar{x}} = \beta_2 \dfrac{\bar{p}}{\bar{x}}$), the average quantity $\bar{x}$ and the average price $\bar{p}$ of the polluting input: $\beta_1 = \bar{x}(1-e)$ and $\beta_2 = e\bar{x}/\bar{p}$. Once the parameters have been estimated, the MEC at E* is

47

given by $\dfrac{\Delta p}{\Delta x} = \dfrac{1}{\beta_2}$ and thus $\Delta p = \dfrac{1}{\beta_2}\Delta x$ where Δp is the MEC associated

with x_s and Δx *is* the reduction of emissions to be achieved in n years.

Once a point on the curve has been estimated we can choose a functional form for the MEC curve and obtain MEC at different levels of emissions.

The two procedures described here are only approximations of the level of some environmental damages, but they have the advantage of being quite easy to implement and to be applied at an aggregate level. In particular, the demand curve for a polluting input is easier to obtain than the corresponding MAC curve.

The accounting treatment of natural resources

As we have seen in section 2, the value of changes in the natural capital stock should be included in our ideal measure of social well-being. The correct measure of NNP should be amended to include:

a)   The value of the flow of renewable resources;
b)   The value of the change in the stock of renewable resources;
c)   The value of the change in the stock of exhaustible resources.

When natural resources are treated like any other capital good, as in Hartwick (1990), the adjustments to national accounts are in rental form, i.e. price minus marginal cost times the flow of the exhaustible resource or the change in the stock of the renewable resource.[27] For renewable resources the level of exploitation valued at its marginal productivity (equal to the resource price in equilibrium) should also be added. The appropriate shadow prices for the depreciation of natural resources can be calculated assuming that the economy follows an optimal path: as we have seen in section 3, we could use optimal prices even if the economy is not at the optimum. These prices will reflect the present value of the future stream of welfare from one more unit of the resource today. They can be multiplied by the corresponding changes in the resources stocks and be added to the conventional national product. Mäler (1996, p. 17) has pointed out some important implications of this way of computing the value of

---

[27]   The change in the stock of a renewable resource is given by the difference between the natural regeneration of the resource in each period of time and the level of exploitation of the resource at the same time.

changes in natural resources stocks. The first one is that changes in the shadow prices are not taken into account, i.e. capital gains or losses are not considered. The second one is that, for exhaustible resources, the change in the stock can be negative unless new discoveries are made. Since new discoveries can be completely random, stock changes should be valued taking into account the likely future discoveries, i.e. contingent claim prices should be calculated, but these may be very difficult to estimate.

For many of the resources included in the stock of natural capital, the same valuation problems arise as with the flow of environmental damages and benefits and the same approaches can be used. However, there are important resources for which there are market prices: oil for instance. Bearing in mind that market prices can be different from optimal prices, there are different ways of computing depreciation of exhaustible resources that can be used in practice.

The Change in Value Method Hartwick & Hageman (1993) have reviewed the approaches to the calculation of exhaustible resource depreciation. Since depreciation is simply the degradation in value of a capital asset under optimal use, it can be computed by calculating the value of the asset, e.g. oil, at the beginning and end of the period and taking the difference, assuming that the resource is being used optimally.

This is called the Change in Value Method. If not available from market data, the value of a pool of oil at any time can be computed by summing the discounted net revenues expected each year for as long as it operates, assuming an optimal schedule of extraction. Let us write the value of an exhaustible resource asset in period t as the sum of the discounted future rents:

(49) $V_t = R_t + 1/(1+r)R_{t+1} + 1/(1+r)^2R_{t+2}+\ldots\ldots\ldots+ 1/(1+r)^nR_{t+n}$

where V is the value of the exhaustible resource stock; R is the annual rent; r is the market interest rate and t+n the time at which the resource will be exhausted. If we begin depletion at year t, n is the years of life of the resource remaining to the pool of oil in year t.

Depreciation, the Change in Value, can be written as:

(50) $V_t - V_{t+1} = R_t - r/(1+r)V_{t+1}$

Computing depreciation by the Change in Value Method usually means estimating the value of the exhaustible resource stock at different time periods, determining the size of the deposits, the future schedule of extraction and the price and costs charged for each ton extracted. Obtaining data on these variables can be a difficult task that can be avoided by deriving an equivalent that is easier to compute.

The total Hotelling rent   The total Hotelling rent, which is easier to compute, has been found to be equivalent to the above definition of depreciation. The total Hotelling rent is the portion of profits that accrues to extraction firms because they are depleting an exhaustible resource. Hotelling's rule implies that if a resource is exhaustible, it will be depleted more slowly than if it were in infinite supply; the resource owner extracts less than the amount that would equate marginal revenue to marginal cost and, consequently, even a competitive firm earns a rent or profit on the marginal ton equal to the difference between the market price and the marginal cost of extraction. The Hotelling rent is therefore a measure of the intertemporal scarcity of the exhaustible resource. It reflects the fact that the exhaustible deposit is shrinking as it is used. The total Hotelling rent (Hotelling rent multiplied by the total quantity extracted) equals depreciation, as proved in Hartwick & Lindsay (1989).

The total Hotelling rent calculation requires much less information that the Change in Value calculation: price (or marginal revenue) and marginal cost to form marginal profits or (Hotelling rent) and the quantity of resource extracted. It is also much easier to compute since it requires no schedule of extraction, no discounting of future receipts and no prediction of future prices or costs.

The main problem in implementing this accounting rule is to obtain marginal extraction costs for the mineral extracted. Usually the problem is solved by using average extraction costs rather than the marginal cost corresponding to the quantity extracted, as in Repetto (1989). However, if the marginal extraction costs curve is increasing, as is usually the case, total extraction costs calculated using average extraction costs are smaller than total extraction costs calculated using marginal costs. In Figure 7, the total extraction cost calculated using the marginal cost at A is given by the area OABE. In this case the area EBCD gives economic depreciation. This area corresponds to what Hartwick (1990) calls "true economic depreciation" given by the formula $[F_r - f_r]R$.

In terms of Figure 7 this becomes (P − MC)OA = CB*EB, the true economic depreciation. If instead of using marginal extraction costs at A, we use average costs, the total extraction cost is given by the area OAB (under the marginal cost schedule), which is smaller than total extraction costs calculated using marginal extraction costs, and economic depreciation is now given by the area OBCD. As Hartwick has pointed out,

50

it is quite clear that the use of average costs overestimates true economic depreciation:

> ... marginal harvesting costs will be difficult to obtain and substituting average harvesting costs will most plausibly over-estimate true economic depreciation (Hartwick 1990, p. 296).

However, when one decides to use marginal extraction costs to calculate true economic depreciation, it can prove difficult to obtain the marginal extraction cost schedule.

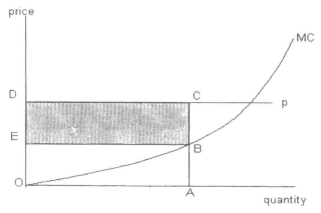

Figure 7: Total extraction costs using marginal and average extraction Costs
Source: Hartwick-Hageman, 1993

The El Serafy method   According to El Serafy (1989) revenues obtained from the exploitation of an exhaustible resource should not be included into NNP, because they are not the result of human production, but the liquidation of an asset. Net receipts from an asset should therefore be divided into two components: the first is capital consumption, i.e. receipts earned at the expense of eroding the value of the resource. This is also called "user cost" or economic depreciation. The other component of net receipts is not earned at the expense of assets' degradation: it is value-added or true income. El Serafy adopts the Hicksian notion of income, i.e. the level of consumption that can be sustained indefinitely. One could sell the resource stock which will be valued according to a finite stream of expected rentals R, place that value, V, in a bank earning an interest X (= rV) every year from then on. Any finite stream of rentals R, earned by the resource stock, can thus be equated to an annuity X every year into the

51

future. The difference between R and X, the sustainable income from the resource, is economic depreciation or "user cost".

(51) $R_t + 1/(1+r)R_{t+1} + 1/(1+r)^2 R_{t+2} + \ldots\ldots + 1/(1+r)^n R_{t+n} = X + 1/(1+r)X + 1/(1+r)^2 X + \ldots\ldots 1/(1+r)^n X + 1/(1+r)^{n+1} X + \ldots\ldots\ldots$

If we assume, as El Serafy does, that the rental R is constant every year we get:

$R_t = R_{t+1} = R_{t+2} = \ldots\ldots\ldots = R_{t+n}$

The finite flow of rentals R earned every year in equal amounts for n+1 years (where n+1 is the last year of resource depletion) is given by the sum of n+1 terms of a geometric progression with reason 1/(1+r) which is equal to:

$$(52) \quad R(1+r)\sum_{t:1}^{n+1} \frac{1}{(1+r)^t} = R\left[\frac{1 - \frac{1}{(1+r)^{n+1}}}{1 - \frac{1}{1+r}}\right] = R\left[\frac{(1+r)^{n+1} - 1}{r(1+r)^n}\right]$$

Expression (52) gives the rent from an exhaustible resource calculated as a finite succession of yearly discounted payments. Similarly, the infinite flow of a constant annuity X is given by the terms of a geometric progression of reason 1/(1+r) and is equal to:

(53) $X + 1/(1+r)X + 1/(1+r)^2 X + \ldots\ldots 1/(1+r)^n X + 1/(1+r)^{n+1} X + \ldots\ldots =$

$$X(1+r)\sum_{t=1}^{\infty} \frac{1}{(1+r)^t} = X\left[\frac{1+r}{r}\right]$$

Setting $\quad R\left[\dfrac{(1+r)^{n+1} - 1}{r(1+r)^n}\right] = X\left[\dfrac{1+r}{r}\right]$, multiplying by $r(1+r)^n$ and

rearranging we obtain economic depreciation or "user cost":

$$(54) \quad R - X = R\left[\frac{1}{(1+r)^{n+1}}\right]$$

Expression (54) is called the El Serafy formula for economic depreciation. The main assumption here is that the yearly receipts from resource extraction are constant, whereas they may in fact change from period to period. In this case the El Serafy formula would be a poor approximation of economic depreciation. Moreover, the accountant still has to guess *n* to compute economic depreciation. El Serafy suggests a procedure that turns the El Serafy formula into the El Serafy method. Here n is calculated by dividing the total reserves remaining by this year's extraction, thus assuming that this year's rate of extraction will continue into the future and that the resource will be extracted until it is physically exhausted. Because of these simplifying assumptions, the El Serafy method is simple to use: one only needs to estimate n, the interest rate and current receipts to estimate economic depreciation, but the resulting depreciation charges are rather arbitrary.

Final considerations

In this review of the main contributions on welfare indices and environmental accounting, it emerges that although this is topical among environmental economists, national accountants and environmentalists, the literature is sometimes unclear on some points. With regards to the theoretical questions, all the models discussed in section 2 share the conclusion that the adjusted NNP is a correct measure of social welfare. However, the models differ in analytical setting and in the way NNP is computed. Weitzman (1976) takes a linear utility function and shows that the current-value Hamiltonian resulting from his maximisation problem is the NNP (as a measure of true welfare). Many authors have tried to extend Weitzman's analysis to the environmental context to determine the adjustments that should be made to NNP to account for the depreciation of environmental resources. However, the results obtained by Weitzman hinge heavily on the specific assumptions of his model.

To overcome this problem, most of theoretical models on environmental accounting resort to some kind of linearisation in order to obtain a current-value Hamiltonian that is linear in its arguments and thus equal to NNP.

Hartwick (1990), for instance, takes a linear approximation to the utility function, whereas Dasgupta, Kriström and Mäler (1997) take a linear approximation to the entire current-value Hamiltonian. In this way they obtain a linear welfare index that is an adjusted NNP without making restricting assumptions on the form of the utility function. Both Hartwick and Dasgupta et al. find that the NNP should be corrected by deducting changes in the stock of natural resources as measured at their shadow values.

However, due to differences in the way the economy is represented and in the functions that they linearise, the correction terms suggested by

Dasgupta et al. differ from those suggested by Hartwick, especially with regards to the treatment of defensive expenditures and changes in environmental amenities.

Defensive expenditures should not be deducted from NNP when they are a determinant of environmental quality and environmental quality is a determinant of welfare, because they affect the level of environmental services and the value of changes in the level of environmental services is included in the welfare measure. A slightly different interpretation emerges from other studies, such as Hamilton (1996). He argues that when household defensive expenditures are a determinant of environmental quality (or services), not only their value should not be deducted from the welfare index, but the adjustment to welfare includes a value greater than the level of households' defensive expenditures. Capital defensive expenditures can be used as an approximation of the depreciation of environmental quality and deducted from NNP. Environmental damages and benefits are difficult to estimates, but the need to obtain estimates at the aggregate levels suggests the use of methodologies that can only be considered approximations to the correct values.

Different methodologies exist for calculating the depreciation of exhaustible resources, here, market prices are available, but the data requirements can be heavy. The most accessible method, in this regard, seems to be the total Hotelling rent.

Two main problems remain. Changes in NNP reflect changes in social welfare when optimal or "local" prices are used. However, the difficulty of estimating accounting prices for all the adjustments proposed means that very few of these adjustments are actually evaluated at accounting prices. We therefore do not know how reliable really is NNP as a measure of welfare.

We have also seen how a big move along the production possibility frontier may lead to a loss of welfare, despite an increase in NNP. This further reduces the reliability of NNP as a measure of welfare in social cost-benefits analysis.

REFERENCES

Atkinson G., Duborg R., Hamilton K., Munasinghe M., Pearce D. (1999) "Measuring Sustainable Development Macroeconomics and the Environment", Edward Elgar Publishers, Cheltenham, UK, (reprinted).
Beltratti, A. (1995) *Environmental Accounting and Policy: a Theoretical Framework*. Manuscript, Fondazione ENI Enrico Mattei, Milan, Italy.

Dasgupta, P., Kriström B. and Mäler K. G. (1997) *The Environment and Net National Product.* In P. Dasgupta and K.G. Mäler eds., "The Environment and Emerging Development Issues", Volume 1, Clarendon Press, Oxford.

El Serafy, S. (1989) *The Proper Calculation of Income from Depletable Natural Resources,* in Y. Ahmad, S. El Serafy and E. Lutz eds., "Environmental Accounting for Sustainable Development". Washington D.C.: World Bank.

Hamilton, K. (1996) *Pollution and Pollution abatement in the National Accounts.* Review of Income and Wealth, Series 42, n. 1, March 1996.

Hartwick, J. (1990) *Natural Resources, National Accounting and Economic Depreciation.* Journal of Public Economics, 43 (1990): 291-304.

Hartwick, J. and A. Hageman (1993) *Economic Depreciation of Mineral Stocks and the contribution of El Serafy,* in E. Lutz eds., "Towards improved accounting for the Environment", Washington D.C., UNSTAT-World Bank Symposium.

Hartwick, J. and R. Lindsay (1989) *NNP and Economic Depreciation of Exhaustible Resource Stock.* Discussion Paper 741. Queen's University, Economics Department, Kingston, Ontario.

Heal, G. (1998) "Valuing the Future: Economic Theory and Sustainability" Columbia University Press, 1998.

Hueting, R. (1980) *New Scarcity and Economic Growth. More Welfare through less Production?* Amsterdam, New York, Oxford: North Holland.

Leipert, C. and U. E. Simonis (1989) *Environmental Protection Expenditures: the German Example.* Rivista Internazionale di Scienze Economiche e Commerciali, vol. 36, 3: 255-270.

Mäler, K. G. (1990) *General Equilibrium Analysis and National Accounting,* in Bojö, Mäler, Unemo eds., "Environment and Development: an economic approach". Kluwer Academic Publisher.

Mäler, K. G. (1991) *National Accounts and Environmental Resources.* Environmental and Resource Economics, 1 (1991): 1-15.

Mäler, K. G. (1995) *Welfare Indices and the Environmental Resource Base,* in Folmer, Gabel and Opschoor eds., "Principles of Environmental and Resource Economics". Edward Elgar.

Mäler, K.G. (1996) *Welfare Indices and the Environmental Resource Base,* mimeo, The Beijer International Institute of Ecological Economics, Stockholm.

Olson, M. (1977) *The Treatment of Externalities in National Income Statistics.* In L. Wingo and A. Evans eds., "Public Economics and the Quality of Life". The Johns Hopkins University Press, Baltimore and London.

Oskam, A. (1993) *External Effects of Agricultural Production in the Netherlands: Environment and Global Warming.* Working Paper, Department of Agricultural Economics and Policy, Wageningen Agricultural University, The Netherlands.

Pemberton, M. and D. Ulph (1998) *Measuring National Income and Measuring Sustainability.* Mimeo, University College London.

Repetto, R., W. McGrath, M. Wells, C. Beer and F. Rossini (1989) *Wasting Assets: Natural Resources in the National Income Accounts.* World Resources Institute, Washington D.C.

UN (1993) *System of Environmental and Economic Accounts.* Studies in Methods, series F, N. 61; Department for Economic and Social Information and Policy Analysis, Statistical Division; United Nations; New York, 1993.

Weitzman, M. (1976) *On the Welfare Significance of National Product in a Dynamic Economy.* Quarterly Journal of Economics, 90: 156-162.

3. Sustainable Development and the Valuation Problem:
Option Values as a Guide-Line for Institutional Choices

Marcello Basili

Introduction

Economic development entails optimal use patterns for resources under a variety of assumptions both about the nature and the objective of the economy in which economic growth occurs. Resources, which are embedded in the economic development, may be classified into three quite different categories: fully reproducible resources (they may be reproduced by human-made capital), renewable resources (there is typically a trade-off between their actual consumption and feasible stocks in the future) and exhaustible resources (there are fixed feasible stocks). Reproducible resources can be used up freely under the marginal productivity rule. Renewable resources are managed as a problem of optimal exploitation for which are defined stocks, natural or technological rates of reproduction and rates of using. Exhaustible resources are exploited by defining an optimal depletion path, given or not some degree of elasticity of substitution between reproducible and non-renewable inputs, induced by technological progress. In fact, an essential[1] natural resource might become inessential by the discovery of a synthetic substitute. Environmental assets are typically both renewable and exhaustible resources and their use entails uncertain consequences and irreversibility in depletion.[2] Uncertainty means that

---

[1] A natural resource is called essential if output of production is zero if it is absent.

[2] Existing theories may diverge about timing and magnitude of consequence, but if a variation in the ecosystem occurs it is supposed to be permanent with respect to the human timescale.

consequences of development decisions cannot be fully determined ex-ante and all the uncontrolled variables of the decision process are random variables and their behavior only depends on the possible state of nature that will occur in the future. Moreover, economic agents have imperfect knowledge that might improve as the time passes.

Irreversibility breaks the temporal symmetry between the past and the future, involving that restoration to an original natural state can be technically impossible or extremely expensive. The intuitive concept of irreversibility like a technological or physical constraint can be generalized to include irreversibility like a restoration cost (the decision to dismantle a dam is an example of costly reversibility).

Since a non-negative change in the stock of natural assets is almost an impossible goal, a development decision has to take into account of both constraints of irreversibility and uncertainty about the values that development would preclude. In this context, preservation of environmental resources provides an extra value, which is irremediably lost, if an irreversible decision is carried out. This extra actual non use-value of preservation can be considered like a form of insurance premium and in such a way it has been related to the concept of option value.

How to evaluate natural assets is a complex question typically because many of these assets are not marketable goods. When an environmental resource is a substitute or a complement of a marketable asset, there are some technical approaches able to capture its use value. However, there is another source of value of natural assets, which is derived from the combination of uncertainty about future benefits embedded in environmental assets and irreversibility in their depletion. In an intertemporal decision process involving environmental resources, it has to consider the possibility of destroying relevant forms of natural capital, at least in a partially irreversible way. The decline in natural habitats is irreversible and the consequences of both natural and man-driven changes are often uncertain in the long term. The interaction among uncertainty, irreversibility of man-driven changes, and the opportunity to postpone investment decisions, spans the concept of environmental option value. Environmental assets (goods and resources) may be considered as (at least in part) irreplaceable assets, the preservation of which has an option value since it leaves open options of consuming (as in the case of a park), production (as in the case of officinal plants), and sometimes of both (as with drinking and irrigation water). The option value of environmental assets is a function of their perceived degree of substitutability by alternative goods, which in turn may change in time with the change of tastes, knowledge concerning their current and potential uses, and availability of new goods. Roughly speaking in the environmental literature option value captures extra benefits related to both conservation and preservation of environmental assets.

In the 1970s, environmentalists have proposed different theories, such as individual's risk attitude, Cicchetti and Freeman (1971) and Schmalensee (1972), to estimate both the future direct use and indirect use value. Models worked out in order to give a characterization of environmental option value gave quite different results because they were focused on different aspects of an investment decision, the more relevant of which were the agents attitude towards risk and the individuals' uncertain future preference with regard to an environmental asset and uncertainty about its future availability. The concept of option is relevant when description of future environmental states of the world is exhaustive and well specified. In this case a more flexible choice commands a sort of risk premium for a risk averse decision-maker because it is a safer store of values (option value).

It has been noticed that the notion of option value refers to a rational behavior under risk, whereas environmental economics should be able to deal with both events that are unanticipated and information about the relative likelihood of events that is imprecise.[3] The occurrence of the virus Ebola, the epidemic infection of BSE (mad cow disease), and the hole in the ozone layer can be considered as unanticipated events. A situation that involves a misspecified and not exhaustive description of environmental states of the world and individual's ambiguous beliefs is referred to as Knightian uncertainty. On the basis of some recent developments in decision theory, which stress the distinction between Knightian risk (soft uncertainty) and Knightian uncertainty (hard uncertainty), in order to take into account of imprecise information, I introduce some procedure to generalize the notions of option value and to update the total economic value of environmental assets. The h-option value depends on the individual's attitude with respect to Knightian uncertainty. The concepts of h-option value is relevant when the decision-maker experiences lack of information due to either a misspecification of the space of environmental states or complete ignorance of the set of the possible future events. In this case, a more flexible choice involves a sort of uncertainty premium for a decision-maker averse to hard uncertainty.

---

[3] Consider either the effect of fire on the Yanomani region of Romania State (Brazil) or the consequence of the Sea Empress oil tanker disaster in South Wales.

As a result, it is possible to derive a new source of value of environmental resources arising from a combination of hard uncertainty in their future use–value and irreversibility in their depletion. This environmental option value is a further correction factor that has to be introduced in the evaluation of total economic value of natural assets.

The plan of the article is as follows: section 2 introduces capacity in order to represent the decision-maker's attitude toward hard uncertainty. In section 3 the concept of h-option value is pointed out. The main results are offered in section 4. Concluding remarks are included in section 5.

Hard uncertainty attitude and capacities

Models explaining option value assume that environmental states of the world have an additive probability of occurring, that is the decision-maker's description of the states of the world is exhaustive. The decision-maker has (explicitly or implicitly) a unique and fully reliable probability distribution over events; moreover she/he possesses an expected utility function linear in probabilities.[4]

Consider a decision problem in which the states of the world included in the model do not exhaust the actual ones. A description of the world is considered as a misspecified model whenever some omitted states are not explicitly included in the model. When the decision-maker does not know how many states are omitted, she/he can represent her/his beliefs by means of a capacity on the set of events.

Let $\Omega=\{\omega_1,...,\omega_n\}$ be a non empty set of states of the world and let $S=2^\Omega$ be the set of all events. The decision-maker chooses an action, technically an act, in the set F. In particular, an act $f \in F$ is a function assigning a consequence to each state, $f:\Omega \to C$, where C is the set of consequences. A function $\mu:S \to R$ is a capacity or a non-additive signed measure if it has the following characteristics:

1. $\mu(\emptyset)=0$, $\mu(\Omega)=1$ (i.e. the capacity is normalized);
2. $\forall A,B \in S$ such that $A \subset B$, $\mu(A) \leq \mu(B)$ (i.e. the capacity is monotone).

---

[4] When the decision-maker "possesses a unique, well-defined classical probability distribution over events, possesses a von Neumann-Morgenstern utility function over outcomes, and ranks subjectively uncertain acts according to the expected utility of their induced probability distributions over outcomes... [she/he] is a probabilistically sophisticated expected utility maximizer" (Machina and Schmeidler, 1992, p.747).

A capacity[5] is convex (concave) if for all $A,B \in S$ such that $A \cup B$, $A \cap B \in S$

$\mu(A \cup B) + \mu(A \cap B) \geq (\leq) \mu(A) + \mu(B)$. It is superadditive (subadditive) if $\mu(A \cup B) \geq (\leq) \mu(A) + \mu(B)$ for all $A,B \in S$ such that $A \cup B \in S$, $A \cap B = \varnothing$.

A real-valued function $f : \Omega \rightarrow R$ is a measurable function if for every $t \in R$, $\{\omega | f(\omega) \geq t\}$ and $\{\omega | f(\omega) > t\}$ are elements of S. Since μ is non-additive, the integration of the real-valued function f with respect to μ is impossible in the Lebesgue sense. The proper integral for non-additive measures is the Choquet integral, originally defined by Choquet (1954) and discussed in Schmeidler (1989). The Choquet integral of f with respect to a capacity μ is defined as

$$\int f d\mu = \int_0^\infty \mu(\{\omega | f(\omega) \geq t\}) dt + \int_{-\infty}^0 \left[\mu(\{\omega | f(\omega) \geq t\}) - \mu(\Omega) \right] dt .$$

The Choquet integral is the standard integral if μ is additive.

The standard subjective expected utility theory (for short SEU) assumes that there exists a bounded real-valued function $u : C \rightarrow R$ and a finitely additive measure or probability p on S, such that expected utility is equal to $\int_\Omega u(f(\omega)) dp$. Moreover u is unique up to an affine transformation.

Given a utility function u, a capacity μ on S and a set of comonotonic[6] acts F, the Choquet integral provides the analog of the SEU and it yields the expected utility of an action with respect to a non-additive measure on events, which has been called Choquet expected utility (for short CEU). The CEU[7] can be represented as a weighted average,[8] where the weights are not given by the non-additive measure but are derived from the cumulative distribution induced by a bijection $\pi : \{\omega_1, ..., \omega_n\} \rightarrow S$, such that $f(\pi_1) \geq ... \geq f(\pi_n)$.

---

[5] It is worth noting that a capacity is additive or a probability measure if for all $A,B \in S$ such that $A \cup B \in S$ and $A \cap B = \varnothing$, $\mu(A \cup B) = \mu(A) + \mu(B)$.

[6] Two acts $f,g \in F$ are comonotonic if and only if there are no $\omega_1, \omega_2 \in \Omega$, so that $f(\omega_1) > f(\omega_2)$ and $g(\omega_1) < g(\omega_2)$. Roughly speaking, two actions are comonotonic if they induce the same ordering of favorable state and the same permutation.

[7] It is worth noting that the preference function defined over acts by the Choquet integral "...is not probabilistically sophisticated, but nevertheless satisfies all the Savage axioms except the Sure-Thing Principle" (Machina and Schmeidler 1992, p.758).

[8] See Wakker 1989.

61

The Choquet integral:

$$\int f d\mu = \int_0^\infty \mu\big(\{\omega|f(\omega) \geq t\}\big)dt + \int_{-\infty}^0 \Big[\mu\big(\{\omega|f(\omega) \geq t\}\big) - \mu(\Omega)\Big]dt \text{ can be}$$

written as $\displaystyle\sum_{k=1}^n f\big(\pi(k)\big)\Big[\mu\big(\{\pi(1),\ldots,\pi(k)\}\big) - \mu\big(\{\pi(1),\ldots,\pi(k-1)\}\big)\Big]$

and the CEU is $\displaystyle\int_\Omega u\big(f(\omega)\big)d\mu$.

The procedure just introduced allows one to represent a preference ordering over the set of the decision maker's feasible acts by a utility function (unique up to a positive linear transformation), even if there are omitted states of the world.

The decision-maker expresses hard uncertainty aversion (preference) if she/he assigns larger probabilities to states when they are unfavorable (favorable), than when they are favorable (unfavorable), that is if her/his non-additive measure is convex (concave).[9] Hence, the convexity (concavity) of the capacity that implies superadditivity (subadditivity) of the Choquet integral[10] captures the decision-maker's attitude toward hard uncertainty.[11]

H-Option value

The decision-maker's behavior is described in a two-period model in which the feasible choices are either to invest at once (irreversible act) or to preserve that is to wait for additional information (reversible act), before investing. In the first period, the decision-maker is uncertain about future states of the world. Uncertainty is resolved by nature in the second period. The decision-maker can choose at the beginning of both periods, but in the

---

[9] Schmeidler (1986) points out that smoothing or averaging utility distribution and the convexity of preference ordering are equivalent to the convexity of the capacity. In fact decision-maker is uncertainty averse if for any three actions $x,y,z \in X$ and $\alpha \in [0,1]$, such that $x \gtrsim z$ and $y \gtrsim z$, then $\alpha x + (1-\alpha)y \geq z$.

[10] If A function $\mu: S \to R$ is a subadditive capacity or a non-additive signed measure, then for functions $f,g \in F$ being

μ -essentially $> -\infty$: $\displaystyle\int_\Omega (f + g)\partial\mu \leq \int_\Omega f\partial\mu + \int_\Omega g\partial\mu$. The corresponding is true for superadditive capacity.

[11] Details are in Chateauneuf 1991.

second period the reversible act is possible if and only if preservation is chosen in the first period. The irreversible act is possible in both periods, that is investment is not a 'now or never' opportunity, but a 'now or next period' one.

Given a capacity μ on S, so that $\mu:S\rightarrow[0,1]$ and consequences express in utility, the quintuple $\{\Omega,S,F,C,\mu\}$ is a decision making problem under uncertainty. In fact, given decision-maker's preference relation $(\geq)$, supposed to satisfy some axioms[12] over actions, there exist both a non-additive measure μ on S and a utility function u, unique up to a positive transformation, such that:

$$f\geq g \text{ iff } \int_{\Omega} u(f(.))dv \geq \int_{\Omega} u(g(.))dv.$$

The decision-maker has preferences over a set of lotteries L. Let L_1 be the subset of the single-stage lotteries[13] and let L_2 the subset of the two-stage lotteries. A two-stage lottery is as follows, in the first stage one event occurs as a result of a 'horse lottery' and in the second stage the prize (technically a consequence) is derived by an objective mechanism such as a roulette wheel or a coin. In the two-stage lottery each stage is clearly distinct and they are separated by a time interval.

The expected utility representation is standard and the decision-maker's expected utility function is continuous and monotone. Choices faced by the decision-maker are acts, which are equivalent to lotteries, and it is assumed that the utility of a lottery is just the expected utility of its prize. Since the consequences are expressed in utility, they may be considered as an expression of welfare changes for the decision-maker, that is a way of measuring the consumer's surplus.

Consider the case of a park which can be either preserved, when the act g is chosen, or destroyed by industrial development, when the act h is chosen, and let net benefit be a linear function of each indivisible act. Option value is defined as the different between the maximum amount the decision-maker would pay to conserve an opportunity for choosing (option price) and her/his expected consumer surplus. Consequence in utility might be considered as an equivalent variation, assuming that if the park is not available in the future, that is irreversible action is made, its price is infinite for any future demand. Moreover, if the decision-maker is certain

---

[12] Schmeidler 1989 sets the properties (axioms) of the decision-maker's preference relation.

[13] More explicitly, $U:L_1\rightarrow R$ represents a weak preference relation $(\geq_l)$ over single-stage lotteries if $\forall\ l^*,l^{**}\in L_1$, it is $l^*\geq_l l^{**}$, if and only if $U(l^*)\geq U(l^{**})$.

about her/his demand to visit the park in the future option price will be equal to her/his consumer surplus. As a consequence, option value is connected to an irreversible act and to an uncertain future demand/supply.

Let states of the world be misspecified, because omitted states of the world are not explicitly included in the model. When the decision-maker does not know how many states are omitted, she/he represents her/his beliefs by a non-additive measure on the events.[14] A situation that involves a misspecified description of states of the world and a non-additive prior has been referred to as hard uncertainty.

If the decision-maker is strictly hard uncertainty averse, she/he assigns a subjective non-additive measure, in the first stage. The hard uncertainty averse decision-maker reveals a kind of pessimism, reflected through a convex capacity, that is she/he gives great weight to possible worse events. The capacity m attached to both uncertain equiprobable events W and Z is $\beta<1/2$, but $m(W \cup Z)=m(\{\omega_1,\omega_2\} \cup \{\omega_3,\omega_4\})= =1>m(W)+m(Z)$ and $m(W \cap Z)=0$. At the end of the first period[15] uncertainty is resolved and in the second period one event in $\{\{\omega_1,\omega_2\},\{\omega_3,\omega_4\}\}$ occurs. The consequence is derived by flipping an unbiased coin. Probabilities s are both symmetric and complementary. Given the set S of all events in Ω, let $m(W)=m(Z)=\beta$ and $s(\omega_1)=s(\omega_2)=1/2$ and $s(\omega_3)=s(\omega_4)=1/2$ be the probabilities and $C=\{1,-1,0,0\}$. The irreversible act h can be represented by a two-stage lottery. Using the CEU[16] the two-stage approach yields $\text{CEU}^{\text{T}}(W)=0$ and $\text{CEU}^{\text{T}}(Z)=0$, then the two-stage lottery can be considered indifferent to a constant act[17] k with consequence 0. In the single-stage formulation, since $v(\omega_1)=(\beta/2)$ and $v(\omega_1,\omega_3,\omega_4)=(\beta)+(\beta/2)$, the CEU of the single-stage lottery is $\text{CEU}^{\text{O}}=(\beta/2)(1)+ +\{1-[(\beta+(\beta/2)]\}(-1)= \beta/2)-(1)+(3\beta/2)=\beta-(1/2)<0$.

Being the decision-maker's hard uncertainty aversion expressed by a convex non-additive measure, a two-stage lottery cannot be reduced to a single-stage lottery by the RCLA.[18] CEU^{O} is the certainty equivalent of the

---

[14] See Gilboa and Schmeidler 1992.

[15] It is worth noting that the assumption β<1/2 only represents non-complementary between probabilities of the events, and it is true for any probability distribution that denies complementary.

[16] The superscript T indicates that CEU is obtained in a two-stage lottery, whereas the superscript O indicates that the CEU is obtained in a single-stage lottery.

[17] An act $k \in X$ is defined to be a constant act if $k(A)=k$, for all $A \in S$.

[18] This framework is derived from Sarin and Wakker (1992). They show that a two-stage lottery and a single-stage lottery yield different results and their outcomes are irreconcilable.

single-stage lottery that is obtained by additivization[19] of the capacity m, calculated as the Choquet integral with respect to $p \in Core(m)$, that is the non-empty core[20] of m on S, such that $p \geq m$ and $\int F_1 \partial m = \min_{p \in core(m)} \int F_1 \partial p$ for $F_1 \in F$, any class of comonotonic functions. CEU^T is the Choquet integral with respect to m of the two-stage lottery, that is the certainty equivalent of the constant act k.

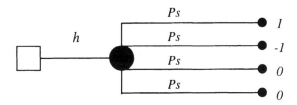

Figure 1: One stage lottery

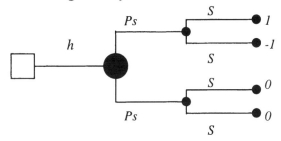

Figure 2: Two stage lottery

If the decision-maker is hard uncertainty seeking he attaches the concave capacity w, in the first stage. A concave capacity characterizes optimism, that is the decision-maker relatively overvalues better events and undervalues the worse ones. The capacity w assigned to both symmetric

---

[19] See Montesano and Giovannoni (1995).

[20] In game theory a convex game is a convex function and every convex game has a non-empty core. The core of m may be interpreted as the set of outcomes, in that no coalition can improve upon any of them. Then, the core(m) is a closed and convex set of finitely additive probability measures that majorize m event-wise (see Gilboa and Schmeidler 1989).

events[21] W and Z is $\gamma > 1/2$, but $w(W \cap Z)=0$ and $w(W \cup Z)=w(\{\omega_1,\omega_2\} \cup \{\omega_3,\omega_4\})=1<w(W)+w(Z)$. If $w(W)=w(Z)=\gamma$, whereas the probability s is unchanged, the $CEU^T(W)=0$ and $CEU^T(Z)=0$ and the two-stage lottery is equivalent to a constant act k' with consequence 0. For coherence with the hard uncertainty averse decision-maker, who holds the prior for the best consequence, in the single-stage lottery $v'(\omega_1)=(\gamma/2)$ and $v'(\omega_1,\omega_3,\omega_4)=(\gamma)+(\gamma/2)$ and $CEU^O=(\gamma/2)(1)+\{1-[(\gamma)+(\gamma/2)]\}(-1)=(\gamma)-(1/2)>0$.

The optimistic decision-maker expects uncertainty to turn out favorably and her/his preference ordering is represented by the CEU calculated with respect to $q \in Q$ (the set of additive measures on S), such that $q \leq w$ and $\int F_2 \partial w = \max_{q \in core(w)} \int F_2 \partial q$ for $F_2 \in F$, any class of comonotonic functions.[22] The RCLA gives the certainty equivalent of the single-stage lottery obtained from w by additivization; meanwhile the two-stage lottery gives the certainty equivalent of the constant act k'.

The difference between CEU^T and CEU^O is called *h-option value*. The sign of this difference depends on the decision-maker's attitude towards hard uncertainty and it is positive (negative) when the decision-maker is hard uncertainty averse (seeking).

Nevertheless the decision-maker can apply the certainty equivalent mechanism (for short CEM) in order to reduce a two-stage lottery to an equivalent single-stage lottery[23] [Figure 3]. In fact, the DM can substitute both sublotteries $W=[1,s;(-1),(1-s)]$ and $Z=[0,s;0,(1-s)]$ with their certainty equivalent $CE(W)$ and $CE(Z)$. By means of the CEM the described two-stage lotteries, with either convex or concave measures in the first stage, can be reduced to equivalent single-stage lotteries. In this way, both of these derived single-stage lotteries can be judged indifferent to a constant act with consequences 0. As result, $L_2\{[(W,\mu);(Z,\mu)]\} \sim L_1\{[(CE(W),\mu);[(CE(Z),\mu)]\}$.

If the capacity in the second stage is non-additive (hard uncertainty is only partially resolved in the first period) the decision-

---

[21] With a concave non-additive measure, non-complementary between events may be represented by assumption $\gamma > 1/2$.

[22] The Choquet integral with respect to a concave capacity w is the maximum among the Lebesgue integrals with respect to a probability $q \in Q$.

[23] See, e.g. Schmeidler 1986.

maker has to evaluate the Choquet Certainty Equivalent[24] of W and Z, that is CCE(W) and CCE(Z). After this evaluating process, once more the decision-maker faces the two equivalent lotteries and $L_2\{[(W,\mu);(Z,\mu)]\}\sim L_1\{[(CCE(W),\mu)];[(CCE(Z),\mu)]\}$.

The CEM produces a general premium that depends on both the DM's hard uncertainty and risk attitude. The decision-maker includes her/his hard uncertainty attitude when she/he evaluates the lottery expected value, then she/he chooses with respect to her/his expected utility function. Since in the standard approach the shape of utility function also expresses the attitude towards risk, the sign of this general uncertainty premium depends on the decision-maker's attitude towards hard and soft uncertainty.

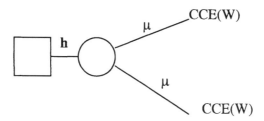

Figure 3: Choquet Certainty Equivalent lottery

Given W and Z, equiprobable and simmetric events, consider two decision-makers with a same utility function (identical risk attitude), but let the former be hard uncertainty neutral and the latter either hard uncertainty averse or seeking.[25]

When the decision-maker's hard uncertainty aversion is expressed by convexity of a capacity m on the set of events, the CEU is at the most equal to the SEU in the core of that capacity, that is $\int_\Omega u(g(.))\partial m = \min \int_\Omega u(g(.))\partial p \mid p \in Core(m)$. The CEU is the minimal

---

[24] The Choquet Certainty Equivalent is equal to the Choquet integral of a sublottery. Then, for every act $f \in F$, $f \sim CCE(f)$. If hard uncertainty is fully resolved in the first period, the CCE coincides with the CE.

[25] Capacities are assumed to be totally monotone that is they are beliefs. If μ is a beliefs, it is non-negative and for every $n \geq 1$ and $A_1,...,A_n \in S$, $\mu(\cup_{i=1}^{n} A_i) \geq \Sigma_{\{I \mid \emptyset \neq I \subseteq \{1,...,n\}\}}(-1)^{|I|+1}\mu(\cap_{i \in I} A_i)$.

expected utility over all probability measures in *Core(m)*, that is the maxmin expected utility with respect to all priors in the core of m.[26] As a result the optimal irreversible act selected by the hard uncertainty averse decision-maker is at the most equal to the optimal irreversible act selected by the hard uncertainty neutral decision-maker.

By contrast, when the hard uncertainty seeking decision-maker is characterized by concavity of a capacity w, $\int_{\Omega} u(g(.))\partial w = \max \int_{\Omega} u(g(.))\partial q \mid q \in Core(w)$ and the CEU is the maximal expected utility over all probability measures in *Core(w)*, that is the maxmax expected utility with respect to all priors in the core of w. [27] As a result the optimal irreversible act selected by the hard uncertainty seeking decision-maker is at the least equal to the optimal irreversible act selected by the hard uncertainty neutral decision-maker.

Main results

The difference ($CEU^T - CEU^O$), that is the difference between the two-stage lottery and the single-stage lottery, depends on the decision-maker's attitude towards hard uncertainty and it may be considered a hard uncertainty premium, called h-option value. Hard uncertainty premium is the maximal amount of money that a pessimistic (optimistic) decision-maker would accept (pay) to bet on a lottery with misspecified states of the world. The sign of the h-option value is positive (negative) when the decision-maker is hard uncertainty averse (seeking), that is ($CEU^T - CEU^O$)>(<)0 if and only if μ is convex (concave).

Once either pessimism or optimism has been evaluated, the decision-maker's behavior depends on his risk attitude. The risk averse decision-maker prefers to get the expected value of a lottery, that is the correspondent constant act or the Choquet expected value, rather than the single-stage lottery, in which prizes are either certainty equivalents or Choquet certainty equivalents. By contrast, the risk seeking decision-maker prefers the single-stage lottery to its Choquet expected value.

---

[26] The Choquet integral of the act f "may be represented as the minimum of all integrals with respect to p ...[and] each of these measures p is simply some 'weight' vector, and the integral of f with respect to p is a p-average over f's values" (Gilboa and Schmeidler 1994).

[27] See Gilboa and Schmeidler 1992.

The generalized option value is the sum of the hard uncertainty premium and the soft uncertainty premium. By means of the Choquet integral, the decision-maker calculates the Choquet expected value of a lottery. By means of the Jensen inequality, the decision-maker defines if $[U(E(l_l))-CEU(l_l)]>(<)0$, as in the standard economics theory.

Proposition 1 *Facing hard uncertainty and an irreversible act, the decision-maker has to evaluate the generalized option value. The sign of the generalized option value depends on the decision-maker's attitude towards hard uncertainty and risk, by means of the Choquet Expected Utility and the Jensen inequality.*

Since $CEU(l_l) \leq SEU(l_l)$, when $m(A) \leq p(A)$ for all $A \in S$, and $SEU(l_l) \geq CEU(l_l)$, when $w(A) \geq q(A)$ for all $A \in S$, the irreversible act chosen by the neutral decision-maker with respect to hard uncertainty is at least (most) equal to the irreversible act chosen by the hard uncertainty averse (seeking) decision-maker.

Proposition 2 *The pessimistic (optimistic) decision-maker shows a preference for a more (less) conservative choice with respect to the hard uncertainty neutral one, because of either incomplete or misspecified description of the states of the world.*

The conclusion that follows from the previous discussion is that option value crucially depends on an individuals' uncertainty aversion and uncertain demand and it is a relevant component of the environmental asset value. Contingent valuation techniques and willingness to pay surveys might give a proxy to option price. In fact, contingent valuation techniques and willingness to pay surveys for changes in environmental resources allocation and in natural resource damage assessment, might give a proxy to option price, or perhaps better, an ex-ante compensating variation (or equivalent variation), and capture hard uncertainty premium.[28]

Concluding remarks

In decision making problems under uncertainty, the decision-maker has to take into account two different situations. First, by attaching a capacity to

---

[28] See Basili and Vercelli 1998.

the events, the decision-maker has to represent her/his knowledge of the world.

Second, the agent has to consider the known economic values and the benefits deriving from a future appreciation of unknown or uncertain resources, which might be precluded by depletion of environmental resources.

This paper points out the existence of the generalized option value, which arises when the decision-maker faces hard uncertainty, which is a usual condition in economic decision making processes involving environmental resources. The generalized option value arises out of uncertainty about consequences of irreversible use of renewable and exhaustible resources and her/his appreciation of those consequences.

Using the certainty equivalent mechanism a two-stage decision process involving environmental resources can be reduced to a single-stage one, so that the decision-maker can evaluate all decisions in the first period. In this case the decision-maker faces an uncertainty premium that can be split into a hard uncertainty premium or h-option value and a soft uncertainty premium or option value. The sign of h-option value depends on the decision-maker's hard uncertainty attitude, which is positive if she/he is a pessimist, by contrast it is an negative if she/he is optimist. The sign of soft uncertainty premium depends on the decision-maker's risk attitude. As a result, the sign of the generalized option value derives from the combination of these premiums.

In benefit-cost analysis of feasible acts, the decision-maker's willingness to pay (to accept) for changes in environmental resources allocation and in natural resource damage assessment, measured by contingent valuation for instance, captures both these premiums. This paper claims that it is possible to split out the generalized option value in the decision-maker's attitude toward risk and reliability of the representation of the states of the world. As a result, for instance in the case of genetically manipulated food, one may distinguish the ecological hazards of genetically engineered crops and reliability of potential hazards.

The paper shows that an irreversible investment made by the hard uncertainty neutral decision-maker is at least (most) equal to the irreversible investment made by the hard uncertainty averse (seeking) one. In fact, the hard uncertainty averse (seeking) decision-maker has a greater (lesser) preference for more reversible actions than the hard uncertainty neutral one. In such a way, *ceteris paribus*, if the decision-maker is hard uncertainty averse, h-option value may be seen as providing a rational policy directed towards a greater preservation of environmental resources, whenever uncertain consequences of using environmental resources might be catastrophic.

REFERENCES

Anscombe, F. J. and R. J. Aumann (1963) 'A Definition of Subjective Probability'. *The Annals of Mathematical Statistics* **34**, 199-205.

Basili, M. and A. Vercelli (1998) 'Environmental Option Values, Uncertainty Aversion and Learning', in G. Chichilnisky, G. Heal and A. Vercelli, ed., *Sustainability: Dynamics and Uncertainty*. Dordrecht: Kluwer.

Choquet, G. (1954) 'Theory of Capacity'. *Annales de l'Institute Fourier* **5**, 131-295.

Cicchetti, C. and A. Freeman (1971) 'Option Demand and Consumer Surplus: Further Comment'. *Quarterly Journal of Economics* **85**, 528-539.

Freeman, A. M. (1984) 'Supply Uncertainty, Option Price, and Option Value'. *Land Economics* **61**, 176-181.

Gilboa, I. (1987) 'Expected Utility Theory with Purely Subjective Non-Additive Probabilities'. *Journal of Mathematical Economics* **16**, 65-88.

Gilboa, I. (1989), 'Additivization of Non-Additive Measures'. *Mathematics of Operations Research* **14**, 1-17.

Gilboa, I. and D. Schmeidler (1989) 'Maxmin Expected Utility with Non-Unique Prior'. *Journal of Mathematical Economics* **18**, 141-153.

Gilboa, I. and D. Schmeidler (1992) 'Canonical Representation of Set Functions'. *Discussion Paper No. 986*, Northwestern University Evanston, Illinois.

Gilboa, I. and D. Schmeidler (1994) 'Additive Representations of Non-Additive Measures and the Choquet Integral'. *Annals of Operations Research* **52**, 43-65.

Giovannoni, F. and A. Montesano (1995) 'Uncertainty Aversion and Aversion to Increasing Uncertainty', in International School of Economic Research, ed., *Decision Under Uncertainty*. University of Siena.

Knight, F. (1921) *Risk, Uncertainty and Profit*. Boston: Houghton Mifflin.

Machina, M. J. (1989) 'Dynamic Consistency and Non-Expected Utility Models of Choice Under Uncertainty'. *Journal of Economic Literature*. **27**, 1622-1668.

Sarin, R. and P. Wakker (1992) 'A Simple Axiomatization of Non-Additive Expected Utility'. *Econometrica*. **60**, 1255-1272.

Savage, L. J. (1954) *The Foundation of Statistics*. New York: Wiley & Sons.

Schmalensee, R. (1972) 'Option Demand and Consumer's Surplus: Valuing Price Changes Under Uncertainty'. *American Economic Review*. **62**, 813-824.

Schmeidler, D. (1986) 'Integral Representation without Additivity'. *Proceedings of the American Mathematical Society*. **97**, 255-261.

Schmeidler, D. (1989) 'Subjective Probability and Expected Utility without Additivity'. *Econometrica*. **57**, 571-587.

Wakker, P. (1989) *Additive Representations of Preferences, a New Foundation of Decision Analysis*. Dordrecht: Kluwer.

PART TWO
THE COMPLEXITY OF INSTITUTIONAL ARRANGEMENTS

4.   "Voluntary" Approaches to Environmental Regulation

Thomas P. Lyon and John W. Maxwell

Introduction

In the 1970s and 1980s most developed nations created a host of new regulations aimed at curbing environmental degradation. The approach taken was typically one of "command and control," which specified in law the standards to be met, often in the form of specific technologies that had to be adopted. Industry fought these new regulations vigorously in many cases, and was repeatedly surprised by the political effectiveness of environmental activists.   From the beginning, command-and-control regulation had been criticized by economists for its costliness and inflexibility, and by the late 1980s market-based instruments for environmental regulation - especially emissions-trading programs - became increasingly common. One of the most striking environmental developments of the 1990s, however, goes beyond even this type of environmental regulation. Under the emerging "voluntary approach" to pollution abatement, firms make commitments to improve their environmental performance above and beyond the level required by law. The purpose of this survey is to make sense of this new and rapidly spreading phenomenon.

We begin, in section 2, with an overview of the types of voluntary programs in existence throughout the United States and Europe. Even a cursory examination reveals that these programs differ significantly amongst themselves, and that the threat of traditional regulation often lurks nearby. In section 3 we present an intellectual framework in which these plans can be understood without ignoring their considerable variety; in the process we survey the nascent theoretical literature on voluntary programs. Section 4 turns to the modest empirical evidence that is available on the efficacy of voluntary programs, while section 5 identifies some key areas for further research and draws together our conclusions.

A survey of voluntary programs

Many terms of art are used to identify voluntary environmental programs, including: business-led environmental strategies, corporate environmentalism, self-regulation, negotiated agreements, environmental covenants, voluntary codes, etc. The European Union Research Network on Market-based Instruments for Sustainable Development offers the following taxonomy of voluntary programs:

1. **Unilateral commitments** by industrial firms. Business-led corporate environmental programs come under this heading.

2. **Public voluntary schemes**, in which participating firms agree to standards that have been developed by public bodies such as environmental agencies.

3. **Negotiated agreements** created out of a dialogue between government authorities and industry, typically containing a target and a timetable for reaching that target. Such agreements may take on the status of legally binding contracts if legislation empowers executive branches of government to sign them.

We describe these different types of programs in more detail below, and then provide brief overviews of programs that illustrate each type in the above taxonomy.

A taxonomy of voluntary programs

Unilateral commitments Many companies and trade associations have initiated environmental programs in recent years. Examples include the Chemical Manufacturers Association's "Responsible Care" program for reducing chemical hazards, Arco's voluntary introduction of reformulated gasoline, McDonalds' replacement of its styrofoam "clamshell" containers with paper packaging, and the German Industry and Trade Association's plan to reduce carbon dioxide emissions.[1] While these programs may be developed after consultation with government bodies, the initiative for the programs rests with industry itself.

Public voluntary schemes Self-regulation in the form of voluntary environmental agreements has a much longer history in Europe than in North America. Furthermore, according to the 1997 report on environmental agreement by the European Environmental Agency (EEA 1997) the number of voluntary environmental agreements in Europe has been rising steadily since the late 1980s.[2] The report notes that there has been little critical assessment of the effectiveness of European voluntary agreements mainly due to the lack of reliable monitoring data and consistent reporting. The 1997 report was aimed in part to fill that gap. The general conclusion of the report concerning voluntary agreements was that they can be an effective tool in complementing existing regulations, and are most effective when they are used to raise awareness of environmental initiatives and provide a forum for information sharing among various parties. However the report was critical of the fact that many voluntary agreements do not include monitoring and reporting requirements. Clearly, as the report points out, this damages the credibility of the voluntary agreements since it does not allow for accountability, and makes ex post evaluation of the effectiveness of the agreements difficult.

Narrowing the definition of voluntary agreements to include "only those commitments undertaken by firms and sector associations which are the result of negotiations with public authorities, or are explicitly recognized by the authorities," the EEA reports that every EU member country had at least one voluntary agreement.[3]

Although the United States has fewer public voluntary schemes than does the EU, such schemes do exist, sponsored by such government agencies as the Environmental Protection Agency (EPA), the Department

---

[1] The first three of these are described in the Appendix. The German program is descrived in the text below.

[2] The bulk of the material presented in this subsection, including our discussions of the operation and effectiveness of specific voluntary agreements in Europe is taken from the 1997 report titled 'Environmental Agreements: Environmental effectiveness' authored by the European Environmental Agency. The interested reader should consult that report for more detail on the examples we discuss in this section, in addition to more examples of voluntary agreements

[3] Note that this definition rules out voluntary initiatives such as trade associations' voluntary codes of conduct.

of the Interior, and the Occupational Safety and Health Administration (OSHA). For example, EPA has sponsored the 33/50 Program (discussed in the text below) and the Green Lights program. The Department of Interior has developed a "No Surprises" policy under which it signs agreements with companies or individual landowners committing not to change the rules applying to a particular piece of property for a fixed period of time. In addition, OSHA has developed a variety of "Voluntary Protection" programs.

Negotiated agreement The two foregoing categories deal with cases where either business or government, respectively, is the prime mover behind a new program. The third category addresses cases where these two actors are both active participants. Not surprisingly, negotiated agreements are more common in Europe, with its tradition of relatively cooperative business/government relations, than in the United States, where adversarial relations between business and government are the norm. European examples include the French agreement on the treatment of end-of-life vehicles (described more fully below), the Swedish agreement of produce responsibility for packaging, and the Dutch policy on implementing target emissions levels in the chemical industry. There are numerous examples in the U.S. as well, including the EPA's Common Sense Initiative and its Project XL.[4]

Three examples

We present below short summaries of three voluntary programs that illustrate the different categories introduced above. A variety of other programs are summarized in the Appendix.

Chemical associations' "Responsible Care" plans: Canada, U.K. and U.S. In 1984 the worst industrial accident in history killed 2500 people and injured 200,000 more when methyl isocyanate gas leaked from a Union Carbide storage tank in Bhopal, India. In 1985 the Canadian Chemical Producers Association established a program called "Responsible Care" designed to improve the industry's environmental and safety performance, and to communicate the industry's improvements to the public. The U.S. Chemical Manufacturers Association and the British Chemical Industries

---

[4]   The Swedish and Dutch cases, as well as the Common Sense Initiative and Project XL, are discussed in the Appendix.

Association followed with similar programs in 1989. According to Frank Whiteley, President of the Chemical Industries Association (CIA), these initiatives were needed to "help regain society's trust". (Fischer and Schot, p. 201) Another key objective, according to industry observers, was "limiting state intervention to a level that is acceptable to the industry". (Fischer and Schot, p. 205.)

Participants in the CIA's Responsible Care program agree to adhere to a set of six guiding principles, all of which aim to ensure that the companies "present an acceptably high level of protection for the health and safety of employees, customers, the public and the environment." (Fischer and Schot, p. 209) The entire CIA membership of roughly 200 companies has signed onto the program, and new firms must accept its principles if they wish to join the association. In the U.S., the industry's self-regulatory activities have become easier for the public to monitor since the creation of the Toxic Release Inventory (TRI), mandated under the Emergency Planning and Right-to-Know Act, Title III of the Superfund Amendments and Reauthorization Act (SARA Title III). In the U.K., however, "[t]he industry is very cautious about the types of data that it makes available" (Fischer and Schot, p. 213.) According to outside observers in the U.K., "[a]t best, the CIA's performance index can be seen as a transitional step toward a much fuller form of disclosure" (Fischer and Schot, p. 213).[5]

The Responsible Care Programs fit quite neatly into the category of unilateral commitments, since the industries were clearly the originators of the programs.

The EPA's 33/50 program for reducing toxic chemical emissions: United States   Starting in 1987, the Environmental Protection Agency (EPA) stepped up its collection of toxics data as a result of Title III of the Superfund Amendments and Reauthorization Act (SARA) of 1986, also known as the Emergency Planning and Community Right-to-Know Act (EPCRA). This law mandates that companies report releases of over 400 different toxic chemicals, many of which are otherwise unregulated. It applies to all manufacturing facilities that have 10 or more employees and that manufacture or process more than 25,000 pounds or use more than 10,000 pounds of any of the reportable chemicals. The EPA makes this

---

[5]   The above account is taken from Peter Simmons and Brian Wynne, 'Responsible Care: Trust, Credibility and Environmental Management,' in Kurt Fischer and Johan Schot, editors, Environmental Strategies for Industry, Washington DC: Island Press, 1993.

information available to the public through the Toxic Release Inventory (TRI). The first year for which data are available is 1987; this information was released to the public in June of 1989.

The availability of the TRI data supported the use of a new voluntary regulatory strategy by the EPA. In February 1991, the EPA announced the"33/50 Program," a voluntary scheme designed to induce firms to cut their emissions of 17 key toxic chemicals 33% by 1992 and 50% by 1995, relative to a 1988 baseline, by providing some favorable publicity and some limited technical assistance. The EPA has been criticized for the program's weak incentives (there are no penalties for failure to participate or failure to achieve the stated goals), and for overstating its results. Nevertheless, the existence of the program may have signalled an increased threat of federal regulation for these chemicals; at any rate, emissions of the 33/50 chemicals fell 42% from 1991 to 1994, while emissions of all other TRI chemicals fell only 22%.

The 33/50 Program fits quite neatly into the category of public voluntary schemes, since the EPA was clearly the originator of the program.

Agreement on the treatment of end-of-life vehicles: France This program, started in 1993, had the goal of reducing the total weight of each automobile destined for a landfill site to be less than 15% of the original weight by 2002 (with an ultimate goal of 5%). From 2002, new model must allow 90% recovery, reuse, or recycling. The participants in drawing up the agreement were the French Ministries of Industry and the Environment, 2 French car manufacturers, 12 French automobile importers, and 8 trade associations covering dismantalers, shredders and recyclers, material producers and equipment suppliers. The agreement carried no explicit sanctions for failure to meet the targets but there was an implicit threat of legislation should there be a failure to meet the program's goals.

Two important initiatives influenced the negotiation of the this voluntary agreement. The first was the EC's (European Commission's) Priority Waste Streams Work on end-of-life vehicles (ELVs), and the threat of stringent legislation concerning such vehicles in Germany. The French government took the lead in coordinating the EC's Priority Waste Streams work. This involved collection and sharing of information on the issue of ELVs at the EU level. While the work on priority waste streams was being undertaken there were moves in Germany to introduce regulation on ELVs. By 1992 this legislative threat by the German Environment Ministry was introduced in draft form.

According to the EEA report "The French Environment Ministry, in an attempt to pre-empt German legislation, began preparing a decree on ELVs. The threat of legislation, and the need to address the threat posed by the German approach led car manufacturers to push for discussions on a [voluntary agreement] with the French government".

One major problem with voluntary agreements highlighted by this case involved monitoring and reporting. Initially the agreement was being monitored by a committee, made up representatives of the agreement's signatories. After criticism over lack of public scrutiny a number of NGOs were invited to attend certain meetings as observers. Initially monitoring provisions were quite poor, with the 1996 progress report containing little comprehensive quantitative data. There are now attempts to improve monitoring and reporting by including progress indicators such as the proportion of ELVs being recycled and the recyclability of new models. To date these provision have not yielded any comprehensive quantitative information.

The French program is a good example of a negotiated agreement, with the government a very active player, but with industry taking an active role as well.

Alternative theories of self-regulation

Corporate environmental initiatives have been attributed to a variety of different motives, including cost-cutting, marketing to "green" consumers willing to pay extra for environmentally-friendly products, and pre-empting government regulation. Understanding what really motivates corporate environmentalism is important for policymakers, since the effectiveness of government environmental policies depends in large part on how corporations will respond to them. It is also important for businesspeople thinking about jumping on the environmental bandwagon, for otherwise they may not obtain the results they hope for. In this section we discuss three general "models" of corporate environmental activity, and consider the social welfare implications of corporate actions within each model. In addition, we attempt to identify the empirical implications of each model, setting up a framework within which we will survey the empirical literature on voluntary environmental activity in the next section. The three hypotheses we consider are:

1. Corporations may have improved the efficiency of their manufacturing processes, generating environmental improvements as a happy byproduct.
2. Shifts in the demand and supply of environmentally friendly products, through the actions of "green" consumers and investors, may be driving changes in corporate behavior.
3. Companies may be taking proactive measures to shape regulatory decisions. They may preempt or mitigate the effects of future government regulations, or alternatively "raise rivals' costs" by signalling that new regulations are inexpensive for firms to implement, and hence should be implemented.

As in many areas of life, more than one of these factors may play a role in motivating corporate behavior in any given situation. We discuss each of them in turn.

Improving corporate productivity

Many writers have argued that companies can cut their costs and improve their environmental performance simultaneously by improving the efficiency of manufacturing processes.[6] Probably the most frequently cited example is 3M Corporation's "Pollution Prevention Pays" program, begun in 1975. For the first time, line workers were involved in identifying opportunities for waste reduction, and between 1975 and 1990, 3M cuts its total emissions of pollution by 50% (530,000 tons). At the same time, the company claims to have saved over $500 million by cutting the costs of raw material, compliance, disposal and liability. Results of this sort, if replicable by other firms, provide support for a "win-win" perspective in which environmental performance and corporate profits walk hand in hand.

From an economic perspective, of course, the puzzle is why there should be any sudden surge in opportunities for making money by cleaning up. Were companies previously sloppy in ignoring internal profit-making opportunities?

Has technological change presented new opportunities for reducing the use of raw materials, thereby lowering costs and environmental

---

[6] Smart (1992) provides a number of examples along these lines. Groenewegen *et al.* offer an extensive guide to the literature on corporate environmental activity.

effluents simultaneously? Has global competition intensified to the point where formerly marginal projects now look attractive? Have workers' attitudes shifted, so that employee morale now depends heavily on corporate environmental performance? None of these questions, to our knowledge, has been satisfactorily answered. In fact, Walley and Whitehead (1994) argue that the opportunities for painless pollution prevention are rapidly being exhausted. Furthermore, Boyd (1998) offers in-depth case studies of three widely-discussed "failures" to adopt pollution prevention, and finds that straightforward economic considerations explain all three.

From the perspective of social welfare, of course, corporate actions that lower costs and also improve the environment are surely beneficial.

Responding to "green" consumers and investors

Another common explanation for corporate environmentalism is that there have been shifts on the demand and supply sides of the marketplace that make environmental activity more profitable. Increasing numbers of consumers, at least in the developed nations of the world, have achieved income levels at which they are willing to pay a premium for environmentally-friendly products.[7] Companies want to appeal to these "green" consumers, and to do so are willing to go above and beyond the levels of care required by environmental regulations. Examples of such environmentally friendly products include organic produce, tuna caught with dolphin-safe nets, biodegradable plastic bags, reformulated gasoline, and McDonalds Corporation's substitution of paper wrapping for styrofoam "clamshell" sandwich containers. The basic notion here is that firms can differentiate their products by improving their environmental qualities, and thereby charge a higher price to high-income consumers.

There is a small but growing economic literature that studies theoretical models of the "green" consumer phenomenon. These models typically build on the basic model of "vertical" product differentiation developed by Gabszewicz and Thisse (1979) and Shaked and Sutton (1982).[8] In this setup, consumers have identical preferences, but income

---

[7] The existence of an 'environmental Kuznets curve' showing the connection between income growth and demand for environmental improvement has been well documented. See, for example, Grossman and Kruger (199*).

[8] Vertical differentiation refers to a situation where all consumers agree on what constitutes higher quality, *e.g.*, most consumers agree that a Mercedes-Benz is preferable to

differences lead to differences in the willingness to pay for product quality. In the standard duopolistic model, two firms offer products of different qualities in one market. The firms bear convex quality-dependent fixed costs and compete in qualities and prices in a two-stage industry game. Since greater product differentiation reduces substitutability and price competition, even firms with identical cost functions will offer distinct qualities in the resulting market equilibrium. In fact, a standard characteristic of such models is that the firms - in their attempts to avoid price competition - engage in "too much" differentiation from the perspective of social welfare.

In the typical version of this model, the consumption of a particular product of a certain quality only affects the utility of the individual consumer. This can be interpreted in two ways. Either individual consumption does not create any externalities, or externalities created by consumption are fully internalized in the utility of the buyer. Some products, such as professional services, generate few externalities. Products involving safety or environmental standards, in contrast, typically create at least some external impacts, and here the assumption of full internalization is less natural, so care must be taken in interpreting the models' results.

Arora and Gangopadhyay (1995) study a standard vertical product quality model where higher quality attracts high-income consumers who internalize the benefits of environmentally friendly products. They replicate some of Ronnen's (1991) results on minimum quality standards, in particular the notion that a minimum quality standard (MQS), if set appropriately, is welfare-improving. The rationale for a standard is that without a standard the firms, in their zeal to avoid price competition, engage in excessive product differentiation. An MQS forces up the quality of the low-quality firm, making it a closer substitute for the product of the high-quality firm, and intensifying price competition. While the high-quality firm responds to an MQS by further raising its quality, this is not enough to offset the effect of the standard, and the quality of the two firms' products moves closer together while prices are forced down.[9]

---

a Yugo. Horizontal differentiation, in contrast, refers to a case where consumers differ in their preferences over a set of similar products, *e.g.* some bicyclists prefer mountain bikes while others prefer touring bikes.

[9] Arora and Gangopadhyay also consider the effects of government taxes and tradeable permits, but since our focus is on voluntary approaches to environmental regulation we do not detail those results here.

Bagnoli and Watts (1995) study several vertical differentiation models where environmental friendliness is only partially internalized by consumers, thereby allowing for a formal analysis of externalities in the context of a "green consumer" model. Their focus is on whether unregulated market forces lead to the socially optimal level of environmental friendliness. They find that in some but not most cases unregulated competition for "green" consumers can provide the socially optimal level of the environmental public good. In particular, when the firms compete by setting prices ("Bertrand competition"), the public good is always underprovided. If firms compete by choosing quantities ("Cournot competition"), then for some parameter values the efficient level of public good provision can be obtained, but only if the public good involves the elimination of a detrimental activity (rather than the expansion of a beneficial activity) and only if all costs of provision are fixed costs as opposed to variable costs.

There appear to be two main lessons to be learned from these models. First, a clear and coherent theory can indeed be developed around the notion that firms may voluntarily make their products more environmentally friendly in order to attract high-income "green" consumers. Second, voluntary action by corporations is unlikely to provide socially optimal results. Government action, at least in principle, can improve the workings of the market. This suggests that a full understanding of corporate environmentalism requires models that combine firms' strategic choices of environmental quality with regulators' strategic choices of environmental standards. We turn to models of this sort later in this section. First, however, we consider another marketplace shift that may also be fueling the move toward corporate environmental activity.

In recent years there has been growth in "socially responsible" mutual funds, which avoid investing in companies deemed irresponsible. Examples of stocks that may be avoided are tobacco, firms with high levels of certain types of pollution, nuclear power, etc. "Green investors" who participate in such funds reduce the supply of capital to the excluded firms, raising capital costs to these firms and shifting the supply curve for the industry upwards. As we discuss below, there is some empirical evidence that stock prices do respond to unfavorable news about corporate pollution, so green investors may be an increasingly important factor determining corporate environmental activity. The theoretical issues here are fairly straightforward; the more important question is the empirical one regarding the relative importance of green investors as a driver of corporate environmentalism.

Optimizing corporate regulatory strategy

A third explanation for the rise of corporate environmentalism is that corporations are acting strategically in the political and regulatory arenas to influence the actions of regulators. This is a rich area of inquiry, and a number of different corporate strategies have been identified in the academic literature. We discuss four such strategies here: 1) Preempting tougher government regulations, 2) Weakening forthcoming regulations, in situations where full preemption is impossible, 3) Reducing the extent of monitoring by regulatory agencies, and 4) Signalling regulators to persuade them to raise rivals' costs. We address each of these in turn.

Preempting tougher regulations In this section we discuss three theoretical models in which voluntary environmental improvements serve to preempt mandatory legislative requirements. The three models correspond to the three elements of the taxonomy we introduced in section 2 above.

Unilateral commitments

As environmental regulation has become an established institution over the past 25 years, corporations increasingly have become able to predict the outcomes of future legislative and regulatory battles. Sophisticated corporate strategists can look ahead to the next wave of likely regulations, and attempt to take proactive steps to shape future laws, rather than passively waiting for regulations to be imposed upon them. If they are sharp enough, firms may be able to preempt future regulations altogether by "self-regulating" with just enough stringency to placate environmentalists and head off the demand for government regulation.

Although it is difficult for outside observers to infer the motives for particular corporate programs, there are many corporate actions that could potentially be interpreted as preemptive. America's "Big Three" automakers' Vehicle Recycling Partnership is limited to creating labeling standards for plastic components, and falls well short of the German program of comprehensive automotive disassembly and reuse. Selective cutting in old growth forests is more environmentally friendly than clear cutting, but many argue that such forests should not be cut at all. The "Responsible Care" program initiated by the Chemical Manufacturers Association (CMA) may be used as a rationale for refusing to adopt more stringent environmental practices; according to Barnard (1990, p. 5), at least one CMA member - Union Carbide - has already done just that. Such examples raise questions about how social welfare is affected when self-regulation preempts government action. This concern is likely to become increasingly important, since "win-win" situations - in which pollution prevention raises both corporate profits and consumer well-being - seem increasingly difficult to find.

Maxwell, Lyon and Hackett (1998) present a model of preemptive self-regulation in which political action is costly for consumers to undertake. Individuals must inform themselves of the implications of pollution control for their well-being, and of the efficacy of various feasible policy remedies. Individuals of similar interests must then coordinate on a mutual strategy for gaining political influence. These various costs are collectively referred to as organizing costs. Even after individuals are organized, they must incur expenses to wield political influence, which might be attained through a variety of means, including lobbying activities, election campaign contributions, and tolerated forms of bribery such as revolving-door arrangements, junkets, and honoraria. Costs that are required after consumers are organized are referred to as influence costs or lobbying costs.[10]

The political costs faced by consumers drive a wedge between the consumer benefits of voluntary abatement and the benefits of mandatory abatement, and firms can take advantage of this wedge to preempt regulation. Naturally, if consumers' costs of political action are too high then consumers are effectively "blockaded" from the political process and self-regulation becomes an unnecessary expenditure and will not be observed. As consumer costs of gaining political influence fall, however, the model predicts that corporate self-regulation will intensify. In other words, an increasing threat of government regulation induces firms to voluntarily reduce pollution emissions.[11] The theory predicts that government actions which significantly lower the information costs faced by consumer and environmental groups would thereby increase the threat

---

[10] Firms face similar tasks, but their organizing costs are typically less than those of consumers, since assessing the costs of regulation to the firm is usually much easier than assessing the health and aesthetic benefits to consumers, and the number of firms in an industry is typically very small relative to the number of consumers. Without loss of generality, then, one can normalize firms' costs of organizing to zero, although they still must incur influence costs to persuade politicians to support corporate positions.

[11] In theory, the cost of preemption may be prohibitive if the threat of regulation is too high. Thus, government policies that subsidize consumer involvement in the political process could paradoxically reduce the level of voluntary environmental improvement.

of regulation faced by firms and increase the incentives for self-regulation. We discuss some empirical evidence on this hypothesis in section 4.

Maxwell, Lyon and Hackett (1998) investigate whether voluntary abatement raises social welfare. When consumer organizing costs are high, firms may be able to preempt regulation with a very modest amount of voluntary abatement, which might be much less than would have been imposed by government. Poorer environmental performance, however, must be weighed against the reductions in regulatory and legislative costs when abatement is voluntary rather than mandatory. The authors show that the regulatory savings more than outweigh any reduced environmental performance.[12] The key idea is that giving consumers some abatement for "free" strengthens consumers' position in the influence game, if it is actually played: they obtain a higher level of total abatement at a lower political cost.

Hence, if consumers allow themselves to be preempted, they must be *even better off* than they would have been had they fought the influence game.

Thus, if preemption occurs, one can presume that both firms and consumers are better off than if consumers had fought to impose standards on an industry that undertook no voluntary abatement.[13]

Maxwell, Lyon and Hackett employ a very general model that builds on the Chicago-style model of interest group pressures developed by Stigler (1971), Peltzman (1976) and Becker (1983). Under this approach, the details of political structure are suppressed in order to focus on the incentives of interest groups, who are viewed as the primary actors. This modeling approach is thus particularly well suited to the analysis of business-led environmental initiatives. The approach could readily be

---

[12] This conclusion holds as long as corporate political lobbying does not directly interfere with the production of political influence by consumer groups. Of course, greater corporate lobbying further offsets consumer lobbying, but welfare reductions only arise if corporate lobbying directly and negatively enters the 'production function' by which consumers produce political pressure.

[13] The model of Maxwell, Lyon and Hackett (1998) can be interpreted as a model of entry deterrence, since the firms invest in pollution control to deter 'entry' by consumers into the political influence game. Unlike standard entry deterrence models, however, here the 'fat cat' strategy - under which investment raises the rival's welfare in the event entry occurs - is effective in preempting entry. (See Tirole 1989), p. 325, for a discussion of 'fat cat' and other entry deterrence stratefies.) The reason is that in ordinary deterrence games staying out yields the potential entrant a fixed reservation level of profits. Here, in contrast, consumers' utility of staying out of the influence game rises as firms invest.

extended to consider a world where it is more costly for interest groups to initiate legislation than to influence the actions of regulatory officials. One could then create a model where firms and consumer/environmental groups press for *regulatory* actions that may preempt future *legislation*. All interest groups will then have incentives to preempt legislation whenever possible. The models we discuss next place more emphasis on the separation between legislators and regulators, and are more specifically tailored to study public voluntary schemes or negotiated agreements.

Public voluntary programs

Segerson and Miceli (1998) present a model in which legislation is threatened if corporations do not produce satisfactory results "voluntarily". In their model, a new piece of environmental legislation mandating pollution reductions is forthcoming with probability p 0 (0,1). The welfare-maximizing regulator cannot unilaterally impose new binding regulations, but is delegated authority to offer the firm a voluntary agreement calling for a greater level of pollution reduction; if the firm accepts the offer, the background threat of legislation is assumed to be removed. Both the physical and the transaction costs of compliance on the part of the firm are assumed to be lower under a voluntary agreement; in addition, the transaction costs faced by the regulator are lower under the voluntary agreement. The costs of legislation are not modeled explicitly.

Given the substantial savings associated with voluntary agreements, Segerson and Miceli show that the equilibrium of this game is for the regulator to offer a voluntary agreement and for the firm to accept. Depending on the parameters of the problem, the voluntary agreement may or may not embody the first-best level of abatement. The first-best level is feasible when the probability of legislation is high and voluntary compliance is much cheaper than mandatory compliance. Segerson and Miceli also discuss the case of multiple firms, and show the importance of clear rules regarding what happens if only a subset of all firms sign the agreement. When firms are homogeneous, the regulator obtains better results by requiring all firms to sign the voluntary agreement if the legislative threat is to be removed; otherwise firms have incentives to free-ride and let their rivals be the ones to shoulder the burden of voluntary abatement. When firms have differing costs of abatement, matters are more complex. Requiring all firms to sign may mean that voluntary agreement is never reached because high-cost firms are unwilling to participate. The regulator must craft the terms of the agreement carefully to balance the

89

benefits of getting more firms to participate against the risk that the agreement will be vetoed by high-cost firms.

In the model of Segerson and Miceli, voluntary agreements always improve welfare if firms are homogeneous due to the cost savings they offer.

When firms are heterogeneous, it is possible that voluntary agreements are not socially desirable. On one hand, a voluntary agreement to which only a single firm is party may be less valuable than mandatory requirements imposed on all firms. On the other hand, obtaining a voluntary agreement from all firms - even those with high compliance costs - may require such a weak abatement requirement that again legislation is preferred. In the model of Segerson and Miceli, the regulator is a welfare-maximizer, so welfare-reducing voluntary agreements will not be observed. In a political economic model, however, one can imagine cases where the regulator may have incentives to implement a welfare-reducing voluntary agreement. The model we describe next focuses on just such situations.

Negotiated agreements

Hansen (1996) presents a model in which voluntary agreements involve direct negotiation between industry and a regulatory body, thereby bypassing the legislative process. Like the papers discussed earlier, Hansen's model views voluntary agreements as supported by the threat of mandatory regulation.

However, Hansen's regulator (which he terms Government) and legislator (Congress) differ in the relative weights they place on firm profitability, revenues from a pollution tax, and environmental improvement.

The utility function of Congress is taken to represent social welfare throughout most of the paper, and Government's divergent interests may lead it to take actions that reduce social welfare. Voluntary agreements produce no tax revenue, and compliance with them may be more or less costly than compliance with legislative requirements. Such agreements come about through Nash bargaining between firms and the Government, each of whom aims to achieve higher utility by preempting Congressional legislation.

It is easy to show that if Congress and Government have the same utility functions, then voluntary agreements only improve social welfare if they lower compliance costs. Conversely, if Congress and Government have different preferences it is easy to come up with cases where voluntary agreements reduce welfare. For example, if Government is sufficiently pro-firm then it can gain by weakening environmental standards and lowering tax revenues, while Congress (and society at large) is worse off. This problem can be alleviated by granting Congress the power to veto any voluntary agreement, thereby assuring that such an agreement must improve social welfare.

Hansen also presents an extension of the model in which various interest groups wield extra influence by applying *ex post* "public criticism" to the actors responsible for a decision, the impact of which is assumed to be a linear function of the actual harm suffered by the interest group. In this setup, both political branches have incentives to avoid responsibility by eschewing action and avoiding criticism. Of course the benefits of avoiding responsibility drive a wedge between the utility function of Congress and social welfare.

This opens up new opportunities for welfare-reducing voluntary agreements that weaken environmental standards and lower tax revenue, and allowing Congress veto power over voluntary agreements no longer assures that such agreements will improve welfare. It is easy to see how welfare reductions can result from criticism by corporate interests. Hansen shows that welfare reductions can also follow from environmental group influence. He allows for criticism of both Congressional goals *ex ante* and legislative achievements *ex post*; in this context Congress again benefits from ducking responsibility and welfare may be reduced.

Summary

All three of the foregoing models explain voluntary environmental improvement as an attempt to avoid resorting to the traditional legislative/regulatory process. Furthermore, all of them highlight the cost savings that come about from voluntary actions, through reduced transaction costs and/or compliance costs. In fact, the similarities between the models suggest that the distinctions between the three different types of voluntary programs are rather blurry. The theoretical work suggests that voluntary abatement is welfare-enhancing in many situations, but it also identifies certain situations in which welfare reductions might occur:

1. Corporate lobbying expenditures significantly reduce the marginal effectiveness of lobbying by consumer and environmental groups, *ceteris paribus*. [Maxwell, Lyon and Hackett]
2. Regulators have political objectives that depart from welfare-maximization *and* are delegated the power to preempt legislation by

accepting weak and/or poorly-designed voluntary programs. [Segerson and Miceli, Hansen]

3. Legislators are scared of public criticism, and thereby delegate too much power to regulators who may lack incentives to maximize welfare. [Hansen]

All these situations revolve around the problem of ensuring that politicians are motivated to pursue social welfare. Maxwell, Lyon and Hackett explicitly assume that the legislature is politically motivated by interest group pressures, yet they still find self-regulation to be welfare-improving in a standard interest-group setting. It appears that the primary welfare concern is that Congress may be systematically biased in favor of one particular interest, above and beyond simply responding proportionately to donations and political pressure.[14] Further research is needed on how and when such bias comes about.

Weakening forthcoming regulations   In some cases, preemption of government regulations may be impossible, but the voluntary actions of firms may influence the regulations subsequently set by government. For example, the Clean Air Act Amendments of 1990 identified 189 toxic chemicals which will be subjected to Maximum Available Control Technology (MACT) standards by the year 2000.[15] Because the details of the standards were not specified by Congress, firms may be able to influence the standards that are actually set through their own actions.

Lutz, Lyon and Maxwell (1998) study this type of setting using a model that includes both "green" consumers and a welfare-maximizing government regulator who sets environmental standards. Like Arora and Gangopadhyay (1995) and Bagnoli and Watts (1995), Lutz, Lyon and Maxwell (1998) work with a duopolistic model of vertical product differentiation, in which high-income consumers are willing to pay a premium for high-quality, environmentally-friendly products. As a result, one firm chooses to specialize in providing a cleaner product to attract these high-income consumers. Furthermore, like Ronnen (1991) and Arora and Gangopadhyay, Lutz, Lyon and Maxwell consider the effects of minimum quality standards. Where they depart from previous work is in: 1) modeling explicitly the standards that would be chosen by a welfare-maximizing regulator (as opposed to simply considering comparative

---

[14]   If legislators delegate the power of preemption to regulatory agencies, then the problem of bias emerges at this level instead.

[15]   See US Environmental Protection Agency (1994).

statics on the level of the standard, as in Ronnen), and 2) allowing the high-quality firm to choose its level of environmental friendliness *before* the government sets its standards.

An apparent example of this sort of strategy is ARCO's introduction of reformulated, cleaner, gasolines under the names "EC-1" (introduced September 1, 1989), "EC-Premium" (September 6, 1990) and "EC-X" (July 1991). The new fuels have garnered ARCO very favorable treatment in the press, and were dubbed the "Product of the Year" by Fortune magazine.[16]

Furthermore, according to the Los Angeles Times, the firm's "return on stockholder equity in 1991 was 29.3%, making ARCO the best performer in the oil industry. Its success is due to an unprecedented new-product development strategy, an environmental strategy. ARCO anticipates environmental regulations to gain significant market advantages."[17]

Lodwrick Cook, ARCO's CEO, clearly believes his own claim that, for firms in the 1990s, "the greatest opportunity for competitive advantage will be in leveraging environmentally improved products and services to differentiate themselves from competitors." However, by "breaking ranks with the majors" and introducing EC-X, Cook not only differentiated his product, he also launched a "frontal assault on those advocating the more expensive vehicles that use methanol fuel."[18]

As mentioned earlier, in vertical differentiation models, firms relax price competition by producing products whose quality levels differ greatly from one another. The level of differentiation chosen tends to be excessive from the perspective of social welfare, and regulators can raise welfare with an appropriately chosen minimum quality standard. This standard is binding for the low-quality firm, while the high-quality firm's best response to the standard is to further raise its quality. Lutz, Lyon and Maxwell show how a high-quality firm can profit by wresting first-mover status from the regulator and proactively committing to a quality level before the regulation is promulgated. The firm chooses a quality that is below its best response to the regulator's preferred standard. For any given standard, this lower quality level increases price competition and reduces profits. Because the welfare-maximizing regulator cares about profits as

---

[16] See Schaefer (1993).

[17] See Piasecki (1992).

[18] Ibid.

well as consumer surplus, however, he chooses a lower standard, and as a result welfare falls.[19]

It is noteworthy that the Lutz, Lyon and Maxwell result on welfare reduction holds even when all of the benefits of environmental improvement are internalized by "green" consumers. When one allows for *external* benefits from pollution abatement, the welfare reduction from weakened environmental standards becomes all the more pronounced. Their results thus provide a lower bound on the negative effects of corporate environmentalism in a setting with minimum quality standards. These simple but robust results provide a new, and rather disquieting, perspective on the growing popularity of corporate environmental management activities.

The analysis of Lutz, Lyon and Maxwell cautions against uncritical celebration of corporate environmentalism. It also provides a contrast to the sanguine conclusions of the preceding section, in which preemption appears to be welfare-improving unless Congress is systematically biased toward a particular interest group. The models in the two sections are built on somewhat differing foundations, however, and future research that synthesizes the two approaches would clearly be valuable. In particular, it would be useful to add to the Lutz, Lyon and Maxwell framework a depiction of the regulatory process that is based on interest-group pressures, since in such a setting externalities may motivate non-consumers to engage in costly lobbying to influence the standard-setting process. Including consumer organizing costs would also allow for the possibility of preemption, and would allow for predictions about when voluntary abatement will tend to be aimed at regulatory preemption and when it will be aimed at weakening future regulatory standards.

Reducing regulatory monitoring   A third way corporate environmentalism can affect regulatory policy is by reducing the stringency with which the firm is treated by regulators. Maxwell and Decker (1998) argue that firm may engage in voluntary environmental investments in order to commit to higher levels of compliance with existing regulations, and may in return, win a lower monitoring rate or laxer permitting scrutiny from regulators. They show that a firm can make an irreversible investment to lower its

---

[19]   While the analysis is done assuming a welfare-maximizing regulator, the same general results would hold even if the regulator was primarily concerned about consumer surplus. The high-quality firm's action would require the regulator to lower the standard in order to prevent the low-quality firm from going out of business.

future costs of complying with environmental standards. If the regulator can observe this investment, then it can infer that the firm is less likely to violate the standards in the future, and will naturally pursue a laxer monitoring policy since the returns to monitoring will have been lowered.[20] Decker (1998) extents the theory of voluntary investment developed in Maxwell and Decker (1998) to explicitly include multiple firms. He shows that firms can take observable pro-environmental actions that convince regulators to focus their monitoring and enforcement efforts *on other* firms.

The welfare effects of corporate environmentalism in both Maxwell and Decker and Decker are ambiguous. In Maxwell and Decker, whether voluntary environmental investments are welfare improving, depends on whether the penalties for non-compliance were excessive relative to the actual social damage caused by a violation of the standard. If the penalties are excessive, the firm has an incentive to *overinvest* in compliance capital. Decker obtains a similar result in the multi-firm case. In Decker the potential for overinvestment is exacerbated by the fact that one firm's investment causes the regulator to monitor other firms more intensively. Each firm's investment thus imposes external costs on other firms, and the firms jointly face a "Prisoners' Dilemma" situation in which the actions that are best from each individual firm's perspective produce results that are collectively bad for the industry as a whole.

Encouraging anticompetitive regulations   We turn now to a fourth and final form of corporate environmentalism as strategic response to regulation, namely actions expressly designed to reduce competition. Earlier researchers have argued that firms may have incentives to encourage regulations that raise industry-wide rents or disadvantage competitors.[21] Voluntary environmental protection may play a role in such strategies. For example, regulators are typically uncertain of the costs of a particular new regulation at the time it is imposed. If those costs turn out to be high, small firms may be forced to exit the industry. Conversely, large firms may benefit from the exit of rivals, and may try to convince

---

[20]   Hemphill (1993) argues that 'The implementation of corporate environmental audits and the active amelioration of deficiencies is viewed by federal and state law enforcement agencies in a positive light when they investigate for criminal prosecutions, and may also be helpful in civil and administrative proceedings.' Helland (1998) provides some evidence that firms self-report environmental violations in an attempt to gain more favourable treatment from the EPA.

[21]   See Bartel and Thomas (1987), Maloney and McCormich (1982), and Pashigian (1985).

regulators that industry-wide compliance costs are low, so stronger regulations might provide substantial benefits at fairly low cost. One way to help convince regulators of this point is for a large firm to make an investment in voluntary abatement, in an attempt to signal to regulators that the cost of abatement is low.

Certainly there are anecdotal reports of firm engaging in tactics like those just described to help stifle competition. For example, DuPont's voluntary acceleration of the phaseout of chloroflourocarbons may have encouraged regulators to put additional pressure on other producers of CFCs.

Anticompetitive strategies may be especially likely in the international arena. Cairncross (1992) provides numerous anecdotes describing how one nation's environmental regulation has been another nation's trade barrier.

Despite the potential use of such strategies, it may be difficult for firms to signal low abatement costs effectively. Smart regulators can infer that in such a setting, large firms will have an incentive to promulgate such regulations only when they feel it will disadvantage rivals. "Fooling" the regulator may thus require very sophisticated strategies. Further research in this area needs to be undertaken.

Welfare effects of corporate environmentalism

Our survey of different theories of corporate environmentalism provides many reasons for optimism with some reasons for concern mixed in. The market pressures created by "green" consumers, investors, and employees are legitimate ways for people to express their preferences, and should be presumed to encourage socially beneficial outcomes. Corporate actions that preempt the emergence of stricter legislation will, except in certain perhaps unlikely cases, provide benefits for both consumers and firms. On the other hand, corporate environmental strategies also may serve to manipulate the regulatory process, possibly inducing regulators to weaken standards to the detriment of consumers and possibly inducing regulators to strengthen standards to the point where competition is attentuated and consumer prices rise unnecessarily. The use of environmental regulation as a trade barrier is a particular worry. Determining which of these stories applies in any given situation is an empirical matter, and it is to the empirical evidence on voluntary environmental amelioration that we now turn.

Lessons from empirical studies of self-regulation

The empirical literature on voluntary environmental amelioration, including self-regulation, is sparse. This is understandable for several reasons. First, corporate voluntary actions in the environmental arena are relatively new.

Second, while casual observation suggests that this type of activity is on the rise, it is not yet common. Third, there is a notable paucity of quantitative data due to the fact that many of the existing programs lack data collection and reporting requirements. Finally, there is the familiar econometric problem of the "missing counterfactual," i.e., the lack of quantifiable data on baseline "business as usual" scenarios. Because voluntary actions necessarily preempt "business as usual" - which might have involved legislated environmental improvements - evaluations of any environmental improvements that do take place under voluntary initiatives are difficult. The few papers that undertake quantitative analyses of corporate environmental actions have generally dealt with discrete choice decisions (either to join a government-proposed pollution-reduction program or to adopt an environmental plan) or continuous decisions regarding the extent of pollution abatement.

Our goal in reviewing the existing literature is to draw out the major findings of each paper and to point to the areas of agreement and disagreement among the papers. This exercise serves two purposes. First, the findings we discuss can be checked against findings arising from future empirical research. These cross-checks are extremely important. As we have mentioned, the existing literature has a very narrow focus, due to data limitations. With a couple of exceptions, the existing literature deals exclusively with toxic chemical emissions in the United States. The main empirical findings arising from the current literature may fail to hold when the focus is broadened to include different corporate environmental acts in different nations. Second, unexplained (or contradictory) findings pose challenges to theoretical researchers, and help focus future empirical research on questions of fact that can aid in the development of future government policies, regulations, and self-regulatory programs.

The following subsection provides a brief review of the empirical methodologies used in the papers we survey. The penultimate subsection summarizes the findings on which the literature is in agreement, and discusses the areas in which the studies disagree. The final subsection offers some suggestions for the advancement and focus of future empirical research.

An overview of the papers

Papers dealing with discrete choice decisions are Henriques and Sadorsky (1996), Arora and Cason (1995 and 1996), Khanna and Damon (1998) and

Karamanos (1998). Each paper estimates the following model using standard probit or logit techniques The binary variable takes on the value 1 if the ith firm undertakes the action under investigation (such as joining a voluntary program) and zero otherwise, is a vector of explanatory variables (e.g., firm and industry financial characteristics, firm pollution characteristics) and perhaps management characteristics, and is a random error term with mean zero. In these models and is interpreted as the probability that a firm with characteristics will undertake the action under investigation.

Henriques and Sadorsky (1996) investigate the corporate decision to voluntarily adopt an environmental plan.[22] Data on the dependent variable and most of their independent variables are drawn from the authors' survey of 750 large Canadian corporations. Published financial data on the corporations surveyed are also used as explanatory variables.

Arora and Cason (1995 and 1996) and Khanna and Damon (1998) focus on firms' decisions to join the EPA's 33/50 program.[23,24] Arora and Cason (1995) use firm level data from 300 corporations, including financial data, advertising intensity, R&D intensity, firm size and past toxic releases as explanatory variables. In Arora and Cason (1996) the data set is expanded to more than six thousand firms by using industry level data, rather than firm specific data, for some explanatory variables (e.g., industry means of advertising intensity and R&D expenditures are used). Khanna and Damon (1998) draw their data exclusively from corporations categorised under the SIC classification "Chemical and Allied Products." In addition to using many of the same firm and industry characteristics as Arora and Cason (1995 and 1996), they include membership in industry associations (in this case the Chemical Manufacturers Association) and a measure of "regulatory pressure" as explanatory variables.[25]

---

[22] The authors leave the exact definition of an 'environmental plan' purposely vague. Thus their dependent variable takes on a value of one whenever the surveyed firm responded positively that it had a formal document describing the plan; had presented the plan to shareholders; had presented a plan to employees; had an environmental, health and safety unit; or had a board or management committee dealing with environmental issues.

[23] See section 2.1 for a detailed overview of the 33/50 program.

[24] Khanna and Damon (1998) also examine the determinants of observed toxic emissions reductions. We review the analytical techniques they use in this examination below.

[25] Their regulatory pressure variable is the ratio of (a) the firm's aggregate volume releases of 189 toxic chemicals that will be subject to maximum available control technology standards (by the year 2000) to (b) the firm's total of all TRI chemical releases. Firms with a higher ratio are expected to feel more pressure to reduce toxic emissions.

Karamanos (1998) examines firms' decisions to join voluntarily the US Climate Challenge program. This program is a voluntary initiative by US electric utilities and the US Department of Energy to reduce emissions of greenhouse gases. Karamanos undertakes a probit analysis using data from all major investor-owned utilities in the U.S.(158 firms). His explanatory variables include firm financial variables, firm size, and past environmental performance, and the environmental performance of the state in which the firm operates.

The studies of Maxwell, Lyon and Hackett (1998), Khanna and Damon (1998), Konar and Cohen (1997a 1997b), and Khanna, Quimio, and Bojilova (1998), and move beyond discrete "participation" decisions by firms to examine the factors motivating firm decisions to voluntarily reduce their emissions of pollutants. Each paper focuses on reductions in toxic chemical emissions using data from the EPA's Toxic Releases Inventory (TRI) . The estimation procedures undertaken in each of these papers are variants of ordinary least squares or basic panel data estimation procedures. Konar and Cohen (1997a) is an exception, and their methodology is discussed below.

Maxwell, Lyon and Hackett probe the general hypothesis that firms are motivated to reduce their toxic emissions in order to avoid (or forestall) future regulation.[26] The data they use as explanatory variables include various measures of political pressure collected at the state level, including variables that are likely to reduce the costs of citizen participation in the political process, *e.g.* the percentage of individuals that are members of environmental groups, the median income level in the state, and the percentage of individuals holding a university degree. Since all pressure variables are collected at the state level, toxic emissions data is aggregated to the state level as well. The authors use political pressure variables, as well as control variables (such as value of industry shipments), to examine both the level of toxic emissions and percentage reductions in toxic emissions, using fixed effects regression analyses and linear regression analyses respectively. The authors examine the emissions of 17 key toxic chemicals - those targeted in the EPA's 33/50 Program - by firms in seven highly-polluting industries (i.e., the top 7 industries emitting the 33/50 chemicals).

Khanna and Damon - in addition to their aforementioned analysis of chemical manufacturers' decisions to join the 33/50 program - use standard

---

[26] See section 3.3 above for a description of the theoretical model used by Maxwell, Lyon and Hackett.

linear regression techniques to examine whether 33/50 membership made a difference in the level of chemical reductions undertaken by firms over the period 1991-1993. To test this hypothesis, the authors include as a right-hand side explanatory variable the predicted probability of participation in the 33/50 program from their first-stage estimation (discussed above). Since the authors also include as right-hand side regressors the variables used to predict participation, the paper allows insight into how firm characteristic variables (other than program participation) affect the likelihood that a firm will reduce its toxic emissions.

Konar and Cohen (1997a and 1997b) and Khanna, Quimio, and Bojilova (1998) examine firms' responses to the public provision of information regarding their TRI emissions. Konar and Cohen (1997a) and Khanna Quimio, Bojilova (1998) specifically examine firms' responses to investor pressure in the form of negative abnormal returns experienced the day following the information release.[27] Konar and Cohen (1997b) focus on the financial ability of firms to respond to investor or consumer pressure following the initial release of TRI information. Konar and Cohen (1997a) use a sample of 130 publicly traded companies from various industries, Konar and Cohen (1997b) use a sample of 520 publicly trading firms operating in industries 20-39 of the Standard Industry Classification (SIC) system, and Khanna, Quimio, Bojilova (1998) draw their data from 91 firms in the chemical industry.

Konar and Cohen (1997a) compare the relative environmental performance of two groups of firms, subsequent to the initial release of TRI information. The first group is made up of the 40 firms experiencing the largest negative abnormal returns following the information release ("Top 40"), while the second group is a portfolio of the remaining firms balanced such that the weight assigned to each firm in a particular industry group is equal to the percentage of firms in that group belonging to the Top 40. Khanna, Quimio and Bojilova examine firms' reactions to information releases over the three year period 1989 to 1991 (as measured by abnormal returns for each of the three years) using on-site releases and off-site transfers of toxics over the period 1991-93.[28] The authors use data on sales,

---

[27] Khanna, Quimio, and Bojilova (1998) examine the impact of both positive and negative abnormal returns and successive TRI information release dates in order to examine the impact of repeated information revelation. Responses to the positive abnormal returns were found to be insignificant. We shall discuss firm responses to the negative returns below.

[28] The authors argue for the existence of a two-year lag in reaction time.

along with research and development, as control variables. Konar and Cohen (1997b) use standard linear regression procedures to examine how both the level and percentage reductions of TRI emissions are affected by firm-level explanatory variables, including various financial characteristics, advertising expenditures, industry concentration, and various environmental characteristics such as TRI emissions and the number of superfund sites the firm owns. The authors also include a dummy variable indicating whether the firm was a member of Konar and Cohen's (1997a) Top 40.

Hart and Ahuja (1996) and Konar and Cohen (1998) use standard linear regression techniques to examine whether good (or bad) environmental performance (measured by reductions in TRI emissions) contributes positively (or negatively) to subsequent firm financial performance. Hart and Ahuja examine whether various measures of financial returns (i.e., return on sales, assets and investment) of member firms of the S&P 500 are positively affected by firm reductions in TRI releases. Konar and Cohen (1998) use Tobin's q as a measure of firm value, and examine whether TRI releases and the number of pending environmental law suits diminish firm value. Each paper draws on the existing industrial organization literature on returns as justification for the inclusion of several common control variables (e.g., industry concentration measures, advertising and R&D to sales ratios).[29]

In the following subsection we summarize the main findings of each of these investigations.

Empirical findings

The extent to which firms undertake voluntary efforts in the environmental area is determined by their willingness *and* ability to do so. The *willingness* to undertake corporate voluntary action is affected by many factors, both inside and outside the corporation. The action's direct profitability (*e.g.* adoption of a greener, and less costly, production process) is certainly a consideration.

However, it is likely the indirect effects associated with the action will ultimately determine whether it is profitable. In other words, a costly action may be profitable if it preempts tighter (more costly) regulation, if it aids in lobbying for weaker regulations, or if it attracts favourable publicity

---

[29] See Schmalensee (1989) for a useful overview of the empirical industrial organization literature on firm financial returns.

leading to increased sales. The ultimate impact of these indirect effects is determined by the actions of various internal and external agents, management and employees, government regulators and legislators, environmental groups, industry peers, and concerned citizens. The *ability* to undertake corporate voluntary actions is likely to be constrained by a firm's financial health and its technological know-how.

In the remainder of this section we categorize the major findings of the extant literature, organized according to various factors that may affect the firm's willingness and ability to undertake corporate voluntary actions.

Firm size and R&D expenditures

Many of the empirical studies of corporate voluntary actions include, as explanatory variables, measures of firm size and research and development (R&D) expenditures. Sales, the number of employees, and the value of assets are used in different studies to proxy for firm size. Regardless of the measure of size used, the overwhelming conclusion is:

Finding 1. Larger firms are more likely to undertake voluntary corporate actions Arora and Cason (1995 and 1996), and Khanna and Damon (1998) find statistically significant support for their hypotheses that larger firms were more likely to join the EPA's 33/50 program. Karamanos (1998) finds that larger investor owned utilities were more likely to join the EPA's Climate Challenge program than their smaller counterparts. Konar and Cohen (1997b) found that larger firms were more likely to reduce, in percentage terms, their emissions of TRI chemicals over the period 1989 to 1992.

Explanations for Finding 1 exist in both the formal literature and the popular press. Because of their higher public profiles, larger firms may feel more pressure to act from environmental groups, politicians, regulators and concerned citizens. It is often asserted that the fixed costs associated with environmental compliance are large enough to generate economies of scale that make it relatively cheaper for large firms to comply with regulations. It is reasonable to assume that the same is true for over-compliance. Finally, larger firms may have better access to capital markets and/or may engage in more R&D. Each of these factors implies that larger firms may find voluntary actions less costly, and we review them in more detail below.

102

Several studies control directly for the impact of R&D expenditure on the likelihood, and extent, of corporate voluntary actions. In total, these studies lend weak support for:

Finding 2. Firms with higher R&D intensities are more likely to undertake voluntary environmental initiatives   Khanna, Quimio, and Bojilova (1998) find weak support for the proposition that greater R&D intensities lead to a reduction in TRI chemical on-site releases and off-site transfers. Arora and Cason (1996) find that firms from industries with greater mean R&D spending intensities were more likely to join the 33/50 program. Khanna and Damon, however, found that within the chemical industry R&D intensity was not a statistically significant factor motivating firms to join the 33/50 program.

Environmental Performance History   How does past environmental performance affect future corporate voluntary actions? On the one hand, poor performance in the past may signal unwillingness or inability to perform well in the future. On the other hand, if environmental performance is measured in percentage-reduction terms (as is often the case) poor past performance may imply lower costs of "performing well" today. In this case, a poor environmental record may actually encourage firms to undertake new voluntary actions, including joining voluntary programs. Finally, firms found to be poor performers are likely to attract the attention of the media and of pressure groups, pushing them towards voluntary action. A careful reading of the empirical literature indicates support for:

Finding 3. Firms with poor environmental records are more likely to participate in voluntary programs   Both studies by Arora and Cason, and the study by Khanna and Damon find that firms with a history of high toxic emissions were more likely to participate in the EPA's 33/50 program. As we have discussed above there are many possible economically rational interpretations for Finding 3. Finding 4, however, presents an interesting economic puzzle.

Finding 4. In deciding to join voluntary programs, firms do not free-ride on their own past clean-up efforts   Due to the fact that firm facilities vary in size and type of production, many voluntary programs measure corporate environmental performance as percentage reductions in pollutants from a given base year. The 33/50 program used 1988 as its base year, but firms were not invited to participate until 1991. Thus, firms knew at the time of their participation decision how successful they had been to date in reducing their emissions. A natural assumption, then, would be that a firm that had already been successful at reducing its emissions would be more likely to join the program (to enjoy the free publicity). However, both Arora and Cason (1996) and Khanna and Damon find that firms with larger percentage emission reductions, prior to making their participation decisions, were not more likely to participate.

One possible, but as yet untested, explanation for the result may have to do with the fact that, due to the public nature of the TRI, firms' progress in emissions reduction could be tracked by the public on an annual basis. Firms which undertook substantial emission reductions early may have feared bad publicity if they failed to maintain their outstanding performance, and may have felt that joining the program would only heighten the probability of bad publicity.

External

Pressures Firms will enter voluntary environmental agreements and/or engage in acts of corporate environmentalism when the benefits of the act exceed its associated costs. As mentioned earlier, these benefits and costs arise from the actions of players both internal and external to the firm.[30] For example, the existence of a large pool of green consumers may raise the benefits of corporate environmental actions through increased sales. At the same time, a large pool of environmentally concerned citizens may raise the costs of not engaging in corporate environmental action, because they may threaten costly future regulations. Regarding external pressure, extant empirical evidence suggests:

Finding 5. The likelihood, and extent, of corporate voluntary actions are increasing in the perceived level of future government regulation, and the strength of community, environmental, and industry group pressure Firms may face pressure to undertake positive environmental initiatives from citizens, who need not be direct consumers. This pressure would normally arise in the form of group pressure either translated through the political process or by direct lobbying of the firm (perhaps with an implicit threat to follow the legislative path, should the company fail to respond). Maxwell,

---

[30] The existing literature focuses exclusively on external players.

Lyon and Hackett provide evidence that factors facilitating pressure group formation and lower lobbying costs prompt firms to undertake greater reductions in their toxic chemical releases. Using state-level data, they find that median income, the percentage of population holding college degrees, and the percentage of population with membership in an environmental group, all contribute positively to reductions in toxic chemical emissions. These findings are consistent with the hypothesis that wealthy, well-educated citizens provide a credible threat to use the regulatory process if corporations do not clean up voluntarily; they also underscore the importance of environmental groups as agents for environmental change.

Of course, the most common way citizens express their public policy concerns is by voting in local, state and federal elections. The preferences of electors will eventually be reflected through legislation and regulatory enforcement. As we have discussed, firms may undertake voluntary actions in order to preempt threatened regulations, or to shape regulations they view as unavoidable. Karamanos finds that private electric utilities located in US states that exhibit relatively poor environmental performance were more likely to join the EPA's Climate Challenge program. In their analysis of the US chemical industry's participation in the 33/50 program, Khanna and Damon find that large emitters of certain toxic chemicals slated for future regulation were more likely to join the 33/50 program. Each of these results is in accord with a key finding of Henriques and Sadorsky: the motivation firms cite most for their decision to voluntarily adopt an environmental plan was to deal with existing and future regulations.

Finally, Khanna and Damon find evidence that members of the Chemical Manufacturers Association were more likely to join the 33/50 program than their non-member counterparts, even though association members were larger emitters. This suggests that industry groups may be another source of pressure on member firms to undertake corporate voluntary actions. Association members may pressure each other because coordinated action may be needed to forestall threatened regulations, or it may aid in shaping the form of future regulations and enforcement procedures. Alternatively, poor performance by one association member may be seen to reflect poorly on all members.

As we have discussed previously, green consumerism is a frequently cited motivation for corporate environmental actions. However, the extant empirical literature yields no consensus regarding the impact of green consumerism. Arora and Cason (1996) and Khanna and Damon find some evidence of green consumer pressure. Specifically, Arora and Cason

(1996) find that firms operating in *industries* with higher advertising to sales ratios were more likely to join the 33/50 program, while Khanna and Damon find, within the chemical industry, that final - good producers were more likely to join the 33/50 program than were their intermediate - good producing counterparts. Konar and Cohen (1997b), however, could not find support for the hypothesis that heavier advertisers undertook greater emission reductions, after controlling for firm size. Furthermore, Arora and Cason (1995) fail to find support for the hypothesis that *firms* with greater advertising to sales ratios were more likely to join the 33/50 program. Finally, when examining actual releases, Khanna and Damon find no support for the hypothesis that producers of final goods in the chemical industry exhibited significantly greater reductions in their emissions of TRI chemicals than firms not engaged in final - good production.

Investor Pressure and Financial Characteristics

A major source of external pressure on the firm not discussed in Finding 5 is investor pressure. In an environmental context we can place investors into one of two categories. The first type, "green" investors, may pressure the firm to sacrifice financial returns in order to enhance its environmental performance. The second type, "traditional" investors, would tend to shun such activity. They would instead pressure the firm to invest in acts of corporate environmentalism only when those acts yield positive financial returns, or help to minimize unavoidable negative returns. Green investors still represent a very small proportion of total investors. It is fruitful, therefore, to look for the effects traditional investors may have on corporate environmental actions.

Is it reasonable to assume that the traditional investor views corporate environmental actions favourably? Several papers study this issue by first examining investor responses to the release of TRI information to the public, and then asking whether firms respond to the signals investors sent them. To gauge investor responses, one could ask whether large emitters of TRI chemicals suffer in the stock market when information of their emissions is released to the public. There is ample evidence that the largest emitters of toxic chemicals did *not* suffer most from the TRI information release. Both Konar and Cohen (1997a) and Khanna, Quimio, and Bojilova (1998) find that firms suffering the greatest negative abnormal stock returns on the day after TRI information was released were not the largest emitters of toxic chemicals.

However, as both papers point out, what should matter to investors is not the level of TRI emissions *per se*, but the difference between the *actual* and the *expected* levels of emissions. The work of Hamilton (1995), Konar and Cohen (1997a) and Khanna, Quimio, and Bojilova (1998) allows one to assert:

Finding 6. Investors react negatively to information revelation regarding higher than expected levels of toxic emissions Both Hamilton (1995), and Konar and Cohen (1997a) find statistically significant evidence of negative abnormal returns for firms with large (but not the largest) emissions of toxic chemicals on the TRI information release day (or the following day). An obvious interpretation of these results is that the largest emitters were expected to be large emitters, so the TRI data provided investors no new information. Further support for this interpretation can be found by noting that in both papers the largest emitters were part of the chemical industry.

Khanna, Quimio, and Bojilova (1998) focus on a sample of 91 firms, all of which are from the chemical industry. They found no evidence of abnormal returns on the first day TRI information was released to the public.

They found, however, that in subsequent years, releases of TRI information did cause poorly performing firms (*i.e.,* those firms who's relative performance in terms of TRI releases worsens over time) to experience negative abnormal returns.

Finding 6 supports the argument that the typical investor views pollution as an economic negative This may occur for several reasons. Investors could link pollution with inefficient production, they could fear that highly-polluting firms will face more intensive regulatory monitoring, or they could fear that large emitters of toxic chemicals face a higher probability of future environmental litigation (even if the releases are legal today). Given these concerns, it is natural to ask whether firms rewarded for improving their environmental performance beyond what is currently required by law. Three papers support the view that:

Finding 7. Firms are rewarded for superior environmental performance Hart and Ahuja (1996) and Khanna, Quimio, and Bojilova (1998) find evidence that voluntary environmental actions lead to negative short-run, but positive long-run, returns. Hart and Ahuja use various measures of financial returns (*i.e.,* returns on assets, sales, and investment) to show that reductions in TRI emissions hurt returns in the year the reductions took place, but had a positive effect on returns in subsequent years. Khanna, Quimio, and

Bojilova show that participation in the 33/50 program lowered chemical firms' current-period return on investment, but increased their market value.[31]   Konar and Cohen (1998) show that poor environmental performance (as measured by TRI releases and the number of pending environmental lawsuits filed against the firm) reduces a firm's intangible asset value.

Another natural question arising from Finding 6 is whether firms take action to counteract environmentally-driven stock price drops. Konar and Cohen (1997a) and Khanna, Quimio, and Bojilova address this question. Their analyses lead to:

Finding 8. Firms respond to environmentally-induced investor pressure by improving their environmental performance   Konar and Cohen (1997a) find that the 40 firms in their sample suffering the largest negative abnormal returns on the first day of TRI information release (the "Top 40"), subsequently improved their environmental performance (as measured by TRI releases and the amount of chemical spills) more than an industry-weighted counter part.   Khanna, Quimio, and Bojilova (1998) find that the firms in their chemical-industry sample, which suffered large abnormal negative returns, subsequently reduced their *releases* of toxic chemicals in favour of off-site transfers.

Summary

The extant empirical literature suggests that large firms undertake voluntary corporate environmental actions for solid economic reasons. Corporations clearly respond to environmentally-driven investor pressures, though we still know little about the underlying drivers of investor concern. There is some evidence that firms respond to threats of future regulations, but the influence of "green consumers" on large manufacturing firms appears to be weak. Thus, the empirical studies overall lend support to the proposition that self-regulation may be effective in yielding environmental improvements *within* the current regulatory system, rather than as a substitute for that system. Furthermore, the current literature provides little evidence that small firms undertake voluntary environmental improvements.   The studies summarized here have focussed almost

---

[31]   The measure they use if 'excess value' as defined by [market value − book value of assets]/sales.

exclusively on larger firms, and there is evidence that the smaller of these firms are less likely to undertake voluntary corporate actions.

Finally, the paucity of empirical work in this area highlights the urgent need for more work in the area. Almost all the existing papers deal with the release of TRI chemicals in the U.S., and only two official programs, 33/50 and Climate Challenge, have been examined. As we stated at the outset, remedying this situation requires greater transparency in the measurement and reporting of the environmental performance of corporations. Indeed, the empirical literature strongly supports the notion that the public release of such information, even absent formal voluntary programs, often prompts firms to improve their environmental performance.

Conclusions

Empirical research shows that superior environmental performance and superior financial performance are intertwined. U.S. financial markets reward firms that go beyond legal mandates for the reduction of toxic emissions, and punish firms that have unexpectedly high levels of toxic releases. Furthermore, firms respond to these financial penalties by improving their environmental performance. Quite simply, voluntary environmental protection appears to make good business sense. What remains unclear, however, are the precise mechanisms that link environmental and financial performance.

While a number of writers argue that pollution reduction cuts costs and improves efficiency, there is no systematic evidence to show how widely this claim might hold true. Nor do the purchasing practices of "green consumers" seem to have much influence on firms' emissions of toxic chemicals, though these consumers may turn out to have an impact on other environmental concerns. There is, however, modest evidence that the threat of future regulation is a significant factor prompting firms to self-regulate; the threat of future legal liability may well serve the same function. The evidence is even clearer that releasing public information about firms' environmental performance spurs them on to greater environmental protection.

Policymakers look with interest on voluntary environmental improvement as a low-cost way to achieve environmental goals. The evidence to date, however, suggests that voluntary activity is a *complement* to regulation, not a substitute, as the threat of regulation is an important factor in motivating corporate voluntary actions. The threat of regulation also helps ensure that voluntary actions will enhance social welfare. Indeed, the theoretical literature generally supports a presumption that self-regulation which preempts government regulation is indeed socially beneficial. Still, there can be cases where self-regulation reduces social welfare. For example, problems can arise when legislation mandates the setting of new standards but defers the details of implementation for an extended period. In this case, firms can weaken the forthcoming standards by voluntarily committing to new technologies offering modest environmental improvements. Once leading firms have adopted new practices, even welfare-maximizing regulators become more reluctant to impose new requirements that will endanger industry profitability. Another concern is that regulators may be delegated excessive powers in cases where they have incentives that diverge from maximizing social welfare. If regulators do not simply balance the political pressures from various interest groups, and instead show systematic bias toward a particular subset of interests, voluntary approaches to environmental protection may fail to serve the public interest.

Despite the foregoing cautions, we believe that voluntary environmental protection is an increasingly important aspect of the corporate landscape, and will produce substantial benefits for both business interests and the public. If nothing else, it provides a way to bypass some costly regulatory proceedings, thereby lowering the transaction costs of achieving society's environmental goals. Further research is important, however, to establish more clearly when self-regulation may be welfare-reducing, and to firm up our understanding of what motivates firms to engage in corporate environmentalism.

Appendix

Agreement of producer responsibility for packaging: Sweden

This agreement (called the REPA scheme) concerns the collection, recycling and material recovery of waste from packaging. The agreement set different percentage targets for re-use or recycling for common packaging materials (e.g., 50% of aluminium and other beverage containers, 95% of standardised glass bottles for beer and soft drinks, 90% of glass bottle for wine and spirits filled in Sweden). These targets were established by the Swedish Ordinance on Producer Responsibility for Packaging, which required they be met by 1997. The agreement began in 1994, has 8,200 corporate signatories and was established through a voluntary industry initiative, recognized by the government. Signatories pay a fee to join REPA and use its recycling and recovery system, or they can set up their own collection and re-use system which will bring them in compliance with the ordinance. The REPA scheme is implemented through the establishment of 5 separate not-for-profit companies which are administered by a general company called Repareistret.

The REPA scheme is actually a voluntary agreement regarding the *implementation* of compliance with the existing Ordinance. Enforcement is essentially set by the Ordinance. The Ordinance requires companies to meet its targets, collect and provide data to the Swedish EPA. The Ordinance is enforced by the municipalities and the EPA.

The REPA scheme was established by representatives of the sectors affected by the Ordinance. The move by the industry to establish REPA was motivated by the legal need to comply with the Ordinance, and the need to control the whole system of collection, re-use and recycling. In particular, the pre-REPA system of collection, re-use and recycling was administered by municipalities, and the firms involved in establishing REPA were concerned about the lack of control they would have over the cost of establishing a larger system that would be required to meet the targets.

The REPA was successful in registering companies which accounted for 85% (by weight) of all packaging used in Sweden. Progress towards targets was mixed: while some targets were exceeded (reusable glass bottles both for beer and spirits, and corrugated paper), progress towards many of the targets was negligible (including aluminum and plastics). In the latter cases, a well developed collection system was not in place prior to the Ordinance, and thus the time line may be viewed as ambitious.[32]

It is difficult to tell whether the establishment of REPA aided in the attainment of the goals of the Ordinance. The companies involved feel that REPA was a more cost-effective means of meeting the goals of the Ordinance. However, some aspects of REPA are troubling. For example, aluminum and steel collection and recycling are done by a single company for one fee, thereby limiting competition between these two metals as sources of packaging.

This case study also highlights the importance of monitoring, enforcement and goal setting in the ultimate success of voluntary

---

[32] Personal communication with the Swedish EPA indicated that the agency was generally satisfied with industry performance, and future targets were being revised to reflect the current progress.

initiatives. In this case the Ordinance *guaranteed* regulatory action if goals were not met.[33]

Declaration of the implementation of environmental policy in the chemical industry: The Netherlands

This declaration - signed by the Association of the Dutch Chemical Industry, 103 individual companies, and various branches of Dutch local and national governments - provides an integrated approach to meeting targets set out in the Integrated Environmental Target Plan (ITEP) for the chemical industry. The ITEP was derived from the National Environmental Policy Plan (NEPP) and its successor (NEPP Plus) developed by the Federal government and published in 1989 (1990 respectively), which set out a strategy aimed at achieving sustainable development by the year 2010.

The NEPP and NEPP Plus contain over 200 *quantified* targets as part of an integrated environmental policy program. The two plans cover a wide range of industries and are aimed at sharing the burden across many sectors of society. The ITEP covers many aspects of the chemical industry and in particular defines targeted reductions (e.g. x% reduction from a base line, usually of 1985) by 1995 with further percentage reductions by 2010, although the 2010 targets are not firm and are currently under review.

The declaration was signed by most major Dutch chemical companies. Some foreign chemical companies operating in the Netherlands (notably US companies) did not sign the declaration, but appear to be altering their operations so as to meet the targets set out in the declaration. Some domestic SMEs also did not sign the declaration, but it is estimated that the declaration covers 97% of total emissions from the Dutch chemical industry.

Each signatory of the declaration is required to submit a Company Environmental Plan-which must be made available to the public - documenting how the company is progressing towards its targets.[34] Also, it is stipulated that signatories must apply best available control technologies. Once the CEP is accepted, it serves as the basis for the company's operating license and in this sense serves as a substitute for the traditional operating license which non-signatories are still required to obtain. The

---

[33] The above account is taken from the European agency's 1998 report on the effectiveness of voluntary environmental agreements, European Environmental Agency (1998).

[34] The CEP must be drawn up every 4 years, and must have an 8 year horizon.

CEP allows the possibility of greater flexibility in operations as long as targets are met. While corporate progress towards targets is set forth in the environmental plan, and self-monitoring is therefore used, the companies are still subject to inspection by regulators who verify the information reported in the CEP.

As of 1997, two years of quantitative data were available. The European Environment Agency (EEA) claims that tangible progress towards stated goals has been made, and it speculates that this progress goes beyond what would have been achieved in the business as usual scenario.[35] Generally speaking, the voluntary agreement is seen as a success, granting companies increased flexibility in their investment planning and consequently inducing costs savings, while at the same time producing tangible progress towards the stipulated environmental goals. Once again, it is clear that target setting, monitoring and reporting are critical to program success.[36]

Declaration by German Industry and Trade Associations on global warming: Germany

The declaration by German Industry and Trade Associations signed in 1995 and extended in 1996 pledged a 20% reduction in CO_2 emissions or specific energy consumption by 2005 (from a 1990 base year) for aggregated industry sectors, with separate targets for each association. Nineteen trade associations signed the 1996 agreement, along with the Ministries of Economics and the Environment, and RWI (Rhine-Westphalian Economics Research Institute) in an independent monitoring role. No individual companies were signatories to the agreement.

The agreement had its origins in early discussions between the German government and industry associations representing the electricity, energy and power, and chemicals industries to develop an agreement to address the climate change problem. The associations represented the main energy suppliers who had the most to lose if the discussed carbon/energy

---

[35] This speculation arises from comparing trends in chemical reductions before the declaration was implemented (1985, 1986, 1989, and 1992) with the two years of available data since the declaration was signed. Given the paucity of data, it is unclear how much environmental improvement can be attributed directly to the declaration.

[36] The above account is taken from the European Environment Agency's 1998 report on the effectiveness of voluntary environmental agreements, European Environment Agency (1998).

tax was to be introduced. As discussions evolved and it became apparent that an agreement could be reached, other parties were brought in, notably the BDI, an association representing most of German industry. The BDI was eventually given overall control of the development and negotiation of the voluntary agreement.

The main motivations for industries to join the voluntary agreement were that signing the agreement would (1) pre-empt a pending heat and waste ordinance, (2) pre-empt the proposed carbon/energy tax, and (3) increase their influence over government policy decisions. Government was interested in entering into the agreement because it wanted an instrument to demonstrate its commitment to meeting national CO_2 targets, while at the same time not increasing regulations on industries already burdened by existing regulation, high energy costs and tax rates (including the unification tax).

In the original 1995 agreement there was no provision for monitoring and reporting. Following public criticism, however, the 1996 agreement did include some monitoring initiatives, including detailed reporting of CO_2 emissions from fossil fuels (to be verified by the RWI).[37] Other criticisms include: 1) the lack of clarity in the agreement about how intra-industry structural change (as opposed to actual energy savings initiatives) will contribute to meeting the targets; 2) the inability to determine how the industry energy saving initiatives differ from those that would be taken under a business-as-usual approach; 3) the absence of publicly negotiated targets; and 4) some provisions in the agreement may give associations added influence over government environmental policy.

While it is still too early to develop a full assessment of the voluntary agreement, Jochem and Eichenhammer (1997) conduct a critical examination of the agreement's targets and are sceptical whether the targets will induce any special efforts on the part of German industry to reduce CO_2 emissions. These authors argue that observed reductions in CO_2 emissions are instead a result of two major exogenous developments: the collapse of the East German Economy following reunification (which significantly reduced power consumption in the east), coupled with the fact that new capital is more energy efficient.[38]

---

[37] The actual reports will be derived from calculations of fuel inputs to the power sector (as gathered by statistical offices) rather than having companies report individually.

[38] The above account is taken from the European Environment Agency's 1998 report on the effectiveness of voluntary environmental agreements, European Environmental Agency (1998).

For years companies have complained that the rigid command-and-control orientation of environmental regulations discourages innovations that might produce better environmental performance at lower cost. Furthermore, the piecemeal development of regulatory policy fosters a lack of coordination across various programs that affect particular industries. The Common Sense Initiative (CSI) and Project XL, proposed by the Bush and Clinton administrations, respectively, aim to support companies in achieving superior environmental performance by granting regulatory "flexibility."

The CSI was motivated out of a desire to coordinate pollution reduction activities across different "media" (land, water and air), and is oriented at the industry level. The pilot phase of the program involved six industries: auto manufacturing, computers and electronics, iron and steel, metal finishing and plating, petroleum refining, and printing. A working group was assembled for each industry, with each group meeting for two-day periods roughly four times per year, usually in Washington, DC. It was hoped that the groups would identify opportunities for regulatory reform and streamlining. Unfortunately, after 16 months of meetings, no regulatory changes have emerged. "By all accounts, CSI has not yet fulfilled its promise to achieve regulatory reform and integration primarily because EPA lacks the statutory authority" to conduct programs on a multi-media basis. (Davies and Mazurek, p. 25)

Project XL, in contrast to the CSI, is oriented toward individual industrial facilities and their surrounding communities. To provide regulatory flexibility, the EPA agreed to a policy of "discretionary enforcement," under which the agency would not pursue statutory violations at participating plants in recognition of their ongoing plans to implement improvements. One significant problem has been that EPA has not been granted legal authority to waive enforcement of any regulations, leaving participating firms vulnerable to third-party enforcement actions. As a result, only three proposals had been granted approval as of January 1998, three years after the program's inception. The first and perhaps most interesting of the permits applies to Intel's "Fab 12" plant near Phoenix, Arizona, the company's newest Pentium fabrication facility. The XL permit allows Intel to make routine changes in production processes without specific authorization from the EPA, as long as the total emissions of conventional and hazardous pollutants do not exceed plant-wide caps. Importantly, the caps are more stringent than required by federal law. The company places great value on XL's regulatory flexibility, for even modest delays can be extremely costly in the rapidly-evolving computer industry. While it is too early to assess the effectiveness of the program, its more focused approach may enable it to produce better results than CSI.[39]

McDonalds Waste Reduction Action Plan: United States

In the late 1980's the McDonalds corporation came under fire for the amount of post-consumer waste its product packaging created each day. In order to keep its pre-made products warm, McDonalds used a Styrofoam packaging system referred to as a "clamshell." This packaging system came under particular attack from environmental critics when information about the damaging effects of CFC on the ozone layer were discovered (CFCs were used in the production of the clamshell packaging system). In 1990, McDonalds U.S.A. formed a task force with the Environmental Defense Fund (EDF) to:

1. Establish ways to reduce, reuse and recycle materials used and wastes generated by the McDonalds system;
2. Provide recommendations consistent with McDonalds' business practices and future growth; and
3. Create a model approach to waste reduction for other companies to emulate.

In 1991 McDonalds announced a Comprehensive Waste Reduction Action Plan (CWRAP) aimed at reducing the amount of solid waste the company generated. At the same time the company announced its decision to abandon the clamshell system. The system was replaced by light cardboard box packaging and wax paper packaging.

Updated annually the CWARP now contains "over 100 initiatives, pilot projects and tests to reduce solid waste in all aspects of its business". McDonalds is heavily involved in recycling and reuse of both its packaging and shipping products, and also uses recycled construction and remodeling materials.

---

[39] The above accounts are taken from Davies and Mazurek (1996), and from Boyd, Krupnick, and Mazurek (1998).

REFERENCES

Arora, Seema and Timothy Cason. "An Experiment in Voluntary Environmental Regulation: Participation in EPA's 33/50 Program," *Journal of Environmental Economics and Management*, v. 28, 1995, pp. 271-286.

Arora, Seema and Timothy Cason. "Why do firms Volunteer to Exceed Environmental Regulations? Understanding Participation in the EPA's 33/50 program," *Land Economics*, v. 72 (4), 1996, pp. 413-432.

Arora, Seema and Subhashis Gangopadhyay. "Toward a Theoretical model of Emissions Control," *Journal of Economic Behavior and Organization*, v. 28, December 1995, pp. 289-309.

Bagnoli, Mark and Susan G. Watts. "Ecolabeling: The Private Provision of a Public Good," mimeo, Indiana University, May 1995.

Barnard, Jayne W. "Exxon Collides with the Valdez Principles," *Business and Society Review*, 1990, pp. 32-35.

Bartel, Ann P. and Lacy Glenn Thomas. "Predation through Regulation: The Wage and Profit Effects of the Occupational Safety and Health Administration and the Environmental Protection Agency," *Journal of Law and Economics*, v. 30, 1987, pp. 239-264.

Becker, Gary S. "A Theory of Competition Among Pressure Groups for Political Influence," *The Quarterly Journal of Economics*, August 1983.

Boyd, James. "Searching for the Profit in Pollution Prevention: Case Studies in the Corporate Evaluation of Environmental Opportunities," mimeo, Resources for the Future, April 1998.

Boyd, James, Alan J. Krupnick, and Jan Mazurek. "Intel's XL Permit: A Framework for Evaluation", Discussion Paper 98-11, January 1998, Resources for the Future, Washington, D.C.

Cairncross, Francis. *Costing the Earth*, Cambridge, MA: Harvard Business School Press, 1992.

Clark, Helen, editor. "Developing the Next Generation of The U.S. EPA's 33/50 Program: A Pollution Prevention Research Project," mimeo, Duke University, Durham, N.C, July 1996.

Davies, Terry and Jan Mazurek. *Industry Incentives for Environmental Improvement: Evaluation of U.S. Federal Initiatives*, Washington, D.C. Global Environmental Management Initiative, 1996.

Decker, Christopher S. "Implications of Regulatory Responsiveness to Corporate Environmental Compliance Strategies," working paper, Department of Business Economics and Public Policy, Kelley School of Business, Indiana University, 1998.

Environmental Protection Agency, Office of Air and Radiation. The Clean Air Act Amendments of 1990 Summary Materials, Washington D.C., 1990.

European Environment Agency *Environmental Agreements: Environmental Effectiveness,* Environmental Issues Series, Vol. 1 (3), European Environmental Agency, Copenhagen, Denmark, 1998.

Gabszewicz, J. J. and J.-F. Thisse. "Price Competition, Quality and Income Disparities," Journal of Economic Theory, 1979, Vol. 20, pp. 340-359.

Groenewegen, Peter, Kurt Fischer, Edith G. Jenkins, and Johan Schot. *The Greening of Industry Resource Guide and Bibliography*. Washington, D.C.: Island Press, 1996.

Grossman, Gene M. and Alan B. Kruger, 'Economic Growth and the Environment' *Quarterly Journal of Economics,* Vol. 118 pp. 353-387, 1995

Hall, Bob and Mary Lee Kerr. *1991-92 Green Index: A State-By-State Guide to the Nation's Environmental Health*, Washington D.C.: Island Press, 1991.

Hamilton, James T. "Pollution as News: Media and Stock Market Reactions to the Toxics Release Inventory Data," *Journal of Environmental Economics and Management*, v. 28, 1995, pp. 98-113.

Hansen, Lars Gårn. "Environmental Regulation through Voluntary Agreements," mimeo, Institute of Local Government Studies-Denmark, November 1996.

Hart, Stuart and Gautam Ahuja, " Does it Pay to be Green? An Empirical Examination of the Relationship Between Emission Reduction and Firm Performance," *Business Strategy and the Environment*, 1996, pp. 30-37.

Helland, Eric. "The Enforcement of Pollution Control Laws: Inspections, Violations, and Self-Reporting," *Review of Economics and Statistics,* v. 80, February 1998, pp. 141-153.

Hemphill, Thomas A. "Corporate Environmentalism and Self-Regulation: Keeping Enforcement Agencies at Bay," *Journal of Environmental Regulation*, Winter 1993/94.

Henriques, Irene and Perry Sadorsky. "The Determinants of an Environmentally Responsible Firms: An Empirical Approach," *Journal of Environmental Economics and Management*, 30(3), 381-359, 1995.

Jochem, Eberhard and Wolfgang Eichhammer. "Voluntary Agreements as a Substitute for Regulations and Economic Instruments: Lessons from the German Voluntary Agreements on CO2-reduction," Nota di Lavoro 19.97, Fondazione Eni Enrico Mattei, Milano, Italy, 1997.

Karamanos, Panagiotis "Factors that Affect. Company participation in Voluntary Environmental Agreements" working paper, School of Public and Environmental Affairs, Indiana University, 1998.

Khanna, Madhu and Lisa Damon. "EPA's Voluntary 33/50 Program: Impact on Toxic Releases and Economic Performance of Firms," mimeo, University of Illinois, Department of Agricultural and Consumer Economics, 1998.

Khanna, Madhu, Willma Rose H. Quimio, and Dora Bojilova. "Toxics Release Information: A Policy Tool for Environmental Protection," *Journal of Environmental Economics and Management*, v. 36, 1998, pp. 243-26

Konar, Shameek and Mark A. Cohen. "Information as Regulation: The Effect of Community Right to Know Laws on Toxic Emissions," *Journal of Environmental Economics and Management*, v. 32, 1997a, pp. 109-124.

Konar, Shameek and Mark A. Cohen. "Why do Firms Pollute (and Reduce) Toxic Emissions" working paper, Owen Graduate School of Management, Vanderbilt University. 1997b.

Konar, Shameek and Mark A. Cohen. "Does the market Value Environmental Performance?" working paper, Owen Graduate School of Management, Vanderbilt University. 1998.

Lutz, Stefan, Thomas P. Lyon and John W. Maxwell. "Strategic Quality Choice with Minimum Quality Standards," Discussion Paper No. 1793, Centre for Economic Policy Research, London, January 1988.

Maloney, M. and R. McCormick. "A Positive Theory of Environmental Quality," *Journal of Law and Economics*, v. 25, April 1982, pp. 99-124.

Maxwell, John W., and Christopher Decker. "Voluntary Environmental Investment and Regulatory Flexibility." working paper, Department of Business Economics and Public Policy, Kelley School of Business, Indiana University, 1998.

Maxwell, J. W., Lyon, T. P. and S. C. Hackett. "Self-Regulation and Social Welfare: The Political Economy of Corporate Environmentalism," Nota di Lavoro 55.98, Fondazione Eni Enrico Mattei, Milano, Italy, 1998.

Moraga, J. L. and N. Padr' on. "Pollution Linked to Consumption: A Study of Policy Instruments in an Environmentally Differentiated Oligopoly," Departamento de Economia, Universidad Carlos III de Madrid, Working Paper No. 97-06, 1997.

Motta, M and J.-F. Thisse. "Minimum Quality Standards as an Environmental Policy: Domestic and International Effects," Fondazione Eni Enrico Mattei, Milano, mimeo.

Pashigian, B. Peter. "Environmental Regulation: Whose Self-Interests Are Being Protected?," *Economic Inquiry*, v. 23, October 1985, pp. 551-584.

Peltzman, Sam, "Toward a More General Theory of Regulation", *Journal of Law and Economics*, v. 19, 1976, pp. 211-248.

Piasecki, B. "Good Deeds and Good Numbers," Los Angeles Times, September 20, 1992.

Porter, Michael, "America's Green Strategy," *Scientific American*, April, 1991.

Ronnen, U. "Minimum Quality Standards, Fixed Costs, and Competition," Rand Journal of Economics, 1991, Vol. 22(4), pp. 490-504.

Schaefer, S. "Cleaner Fuels for Competitive Advantage: ARCO and EC-1," Stanford Graduate School of Business Case BE-10, 1993.

Schmalensee, Richard. "Inter-Industry studies of Structure and Performance," in *Handbook of Industrial Organization,* Richard Schmalensee and Robert Willing Eds., North Holland, 1989.

Segerson, Kathleen and Thomas J. Miceli. "Voluntary Approaches to Environmental Protection: The Role of Legislative Threats," *journal of Environmental Economics and Management*, 1998.

Shaked, A. and J. Sutton. "Relaxing Price Competition Through Product Differentiation," Review of Economic Studies, 1982, Vol. 49, pp. 3-13.

Simmons, Peter and Brian Wynne, "Responsible Care: Trust, Credibility and Environmental Management," in Kurt Fischer and Johan Schot, eds., *Environmental Strategies for Industry*, Washington, D.C.: Island Press, 1993.

Smart, Bruce. *Beyond Compliance: A New Industry View of the Environment*, Washington, D.C.: World Resources Institute, 1992.

Stigler, George. "The Theory of Economic Regulation," *Bell Journal of Economics and Management Science*, v. 2, 1971, pp. 3-21.

Tirole, Jean. *The Theory of Industrial Organization*. Cambridge, MA: The MIT Press, 1989.

U.S. Environmental Protection Agency. "EPA's Voluntary Programs: A Summary List Prepared by The 33/50 Program" Washington, DC, June 1994.

Walley, Noah and Bradley Whitehead. "It's Not Easy Being Green," *Harvard Business Review*, May-June, 1994.

5. Economic Institutions and Common-Pool Resources:
The Role of Exit Costs in Enforcing Full Cooperation

Antonio Nicita

Introduction

In the standard "tragedy of the commons", originally formulated by G. Hardin (1968), collective action by rational herders leads to a "tragedy", i.e. to the environmental degradation inevitable when many individuals use a scarce resource without any concern about deployment rates, as herders would do in a common pasture.

In the absence of well defined property rights on common-pool resources (CPRs), each member of the pool is induced to increase his consumption of CPR, receiving a direct benefit and only later suffering the costs related to environmental deterioration.

Traditional analysis of the tragedy of the commons has applied the standard prisoner's dilemma game (PD) to common-pool resources.

According to this perspective, the "tragedy" is depicted by the paradox that individually rational strategies lead to collectively irrational outcomes, a conclusion already envisaged in the work by M. Olson (1965).

As Olson pointed out, in the absence of coercion (i.e. of an external authority able to detect and enforce sanctions on free-riders), the collective action of the members of a group managing a CPR does not have an efficient social outcome and rational self-interested individuals do not maximise their common economic interests. When it is not possible to prevent a member of the group from obtaining the benefits of a collective good, each member will be induced to free-ride, once the good is produced. In the bilateral non-cooperative prisoner's game, with complete information, the dominant strategy of players will be to free-ride, promoting a solution which is a third-best for both players.

Among the solutions and policy prescriptions suggested by a vast scholarly literature,[1] I recall: *(i)* the "coercive" solution (Ophuls, 1973) for which only public control over common-pool resources can align members' behaviours in an efficient way, and *(ii)* the "de-centralised" solution (so-called *"privatisation"*), for which only Coasian re-allocation of property rights over common-pool resources can induce members to reduce the gap between private benefits and collective costs. Unfortunately, these two extreme solutions both require an external authority to impose the desired institutional change on the members' pool, in order to centralise or decentralise resource governance. However, in most "commons games", the "authority failure" is the cause - and not merely the effect - of a common-pool resources regime. In other words, when it is difficult to define and enforce property rights on natural resources, the immediate consequence is the emergence of common-pool or open access resources. Taking fisheries in the Mediterranean Sea as an example: each Mediterranean state would have to employ vast economic resources to monitor, prevent and punish free-riding by other states. The question to be solved is how to define property rights on fisheries when numerous sovereign states exploit resources of a common sea.

Advocating central regulation or privatisation in such a situation is simply unrealistic. In the absence of an international authority, only international negotiations can be used to deter free-riding and tragedies in the international arena. We are therefore at the starting line defined by Hardin: how to govern common-pool resources when exogenous enforcement is not possible?

The aim of this chapter is to answer to the above question by merging two approaches: analysis of free-riding problems in commons frameworks and analysis of hold-up problems in incomplete contract frameworks. The literature on incomplete contracts and that on tragedy of the commons have in common the problem of "authority failure" and the emerging of self-enforcing strategies to deter ex-post opportunism. According to Williamson (1985), to prevent post-contractual opportunism in an incomplete contract framework, economic agents have to design endogenous enforcement devices (*"private orderings"*). Here, we try to explain how strategies that affect agents' exit options may enhance the pursuit of a second best outcome.

---

[1]   For a very stimulating survey of 'tragedies' in CPRs, see E. Ostrom (1990).

Free-riding, hold-up and tragedies

Let us start by describing the commons symmetric game as a prisoner's dilemma game (PD) where two identical agents – say two herders – have to coodinate their actions (cooperate, exploit). Their (symmetric) payoffs are π_i where $\pi_i=(a,b,c,d)$ and $a_i > b_i > c_i > d_i \ \forall \ i=(1,2)$.

Table. 1: Symmetric PD game

		Agent 2	
		Cooperate	*Exploit*
Agent 1	*Cooperate*	$b_1; b_2$	$d_1; a_2$
	Exploit	$a_1; d_2$	$c_1; c_2$

In the standard PD depicted above, the co-operative solution is not reached in a one-shot game. The *Folk theorem* may thus be evoked to sustain (under given conditions) a cooperative solution in the repeated game.

However, in most commons games the strategic context faced by agents is not a traditional PD, but a one-shot *chicken game* (CG) (Carraro and Siniscalco, 1992), where each agent prefers a cooperation policy to a depletion one, but delegates the other agent to implement the cooperative solution. In the CG, we therefore have two Nash equilibria, each characterised by unilateral cooperation.

Table. 2: Symmetric Chicken game

		Agent 2	
		Cooperate	*Exploit*
Agent 1	*Cooperate*	$b_1; b_2$	$c_1; a_2$
	Exploit	$a_1; c_2$	$d_1; d_2$

In a CG setting, to avoid the worst solution (bilateral exploitation) the coordination problem might be formulated as a meta-game (cooperative or non cooperative game). According to this perspective, Carraro and Siniscalco (1992) suggested that in order to modify agents' payoffs and select a unique Nash equilibrium characterised by a sustainable rate of natural resources depletion, the coordination problem may be solved by introducing some appropriate "external" side-payment in the environmental game, or by merging the environmental game with another game, the so-called *linked issue game,* to alter the parties' payoffs. Following this view, some scholars have formalised *linked games* in the form of *mirror image games,* where beside the principal (environmental) game a "mirror game" is built in such a way that agents' asymmetries are solved in the merged game, and both agents are induced to select the cooperative solution in the principal game.

However, for our purposes, the linked game approach does not envisage that in order to be effective, it has to register perfect timing of transfers between agents involved in the principal game. As we know from the literature on post-contractual opportunism in incomplete contracts, it may be not sufficient to transfer side-payments to deter counterpart opportunism, if the opportunist "hits and runs" without suffering any additional cost. This is the well-known "hold-up problem". In the absence of coercion or binding agreements, the opportunist may "wait" to cooperate until the counterpart is committed to cooperate (as in the CG) and then ask for renegotiation of "contractual" terms, failing which the relationship breaks down.

As we can see, the hold-up problem may only be solved by synchronised exchange of "side-payments" and "actions" (investments). However, in a repeated setting, it may be impossible, or very costly, to detect parties' actions and hold-up may always occur.

In the sections that follow we propose a unified framework to analyse free-riding and hold-up problems in strategic situations like the symmetric PD and CG.

We suggest that linked games should be structured so as to increase the exit costs of the potential opportunist. In order to prevent ex-post hold-up, we show that, the linked game (in the form of side-payments) should increase the ex-post specificity of opportunistic agents. For a linked game to lock agents into the contractual relationship, it is thus necessary to implement strategic specific investments, i.e. investments which provide a higher ex-post value than the ex-ante parties' outside options only if transaction takes place.

Once made, a specific investment will lock the investors into the contractual relationship by raising their ex-post exit costs: outside the transaction, the ex-post value of specific assets or investments will thus be lower than their best ex-ante outside options. However, for an incomplete contract, to lock-in the parties involved, every change affecting parties' outside options must be non-binding (MacLeod and Malcomson, 1993). In order to guarantee this outcome, the ex-post division of surplus between the parties should give them a payoff greater than the one provided by their next best alternative (Grossman and Hart, 1986; Hart and Moore, 1990; Hart, 1985). When a linked game increases the specificity of the potential opportunist, commitment is credible and the cooperation of agents in CPRs is suitable even in the absence of third party enforcement. The model developed here clarifies these conclusions.

Let us assume, for instance, a two-agent pool (firm A and firm B) managing a CPR (a lake). Assume that clean water from the lake is an input in the firms' production processes and polluted affluent is discharged into the lake. The water of the lake can be used as an input for as long as its pollution level is below a quality threshold, Q^*. To reduce the pollution of the lake, without decreasing production levels, the agents may use process I, which ensures cleaner water in the lake, when the process operates at optimal level I^*, so that $Q(I^*)>Q^*$. Once the process is adopted, each firm has to decide whether or not to use it, at every productive cycle. Use of the process at optimal level implies fixed costs, so that each agent compares the actual cost of use of the process with future earnings in terms of input availability. We also assume that both the agents use the process. However, while the use of the process can be observed, whether or not it is working at optimal level in a given production cycle cannot be observed.

Firms A and B can thus choose the following strategies:

1. *Commitment (C):* "use the process at its optimal level, whatever the other firm does";
2. *Reciprocity (R):* "use the process at its optimal level only if the other firm is doing the same";
3. *Hold-up (H):* "wait until the other firm uses the process at its optimal level and then renegotiate own use of the process at its optimal level (ex-post costs sharing)";
4. *Free-riding (F):* "do not use the process at its optimal level, whatever the other firm does".

If both agents commit to undertake strategy C, they will achieve the first best solution without any additional cost. However if at least one agent is likely to select strategies R, H or F, some form of institutional arrangement is required to induce firms to cooperate.

Among the arrangements proposed in the related literature on commons and incomplete contracts, we recall the following:

1. firm A may merge into firm B (horizontal integration) (Williamson, 1985; Grossman and Hart, 1986);
2. the firms may reciprocally transfer strategic assets to achieve joint control over each other (*hostage exchange*, Williamson, 1985);
3. the firms may sign a contract with high renegotiation penalties to deter hold-up[2] strategies (MacLeod and Malcomson, 1993);
4. the firms may define *self-enforcing* deterrence strategies to credibly punish deviations from cooperative strategy (*trigger strategies and penal codes* (Abreu, 1988));
5. firms may strategically recur to the sanctioning role of exit costs, to enforce the cooperative solution (Lin, 1991, Putterman and Skillman, 1992; MacLeod, 1993).

In the following sections, we assume that strategies (a) and (b) are not available, focusing then on strategies (c)-(e). Solution (c) requires the existence of a neutral third party, an assumption here excluded in order to analyse also transnational commons. We focus analysis on the effectiveness of solutions (d) and (e). In particular, analysis of (e) helps to clarify the role of exit costs as enforcement devices.

The *commons game* with zero exit options: the one-shot game

Assume that each agent payoff in the above example is given by

$$(1)\ \pi^j (u^j, i^j) = u^j - c^j (i^j) \quad \forall\ j=(A, B)$$

where i^j is the start of the process (we define it as an "investment"), u^j are j's benefits, and $c^j (i^j)$ is the cost function associated with the start of the process, where $c^{\prime j} > 0, c^{\prime\prime j} > 0$, $\forall\ i^j \geq 0$. We assume that $c^j = c^h \ \forall\ j$

---

[2]   In particular, each agent may transfer at the beginning of each period to a third party a side-payment which will be re-paid at the end of each period, only subject to the consent of the other firm.

and $h = (A, B)$ with $j \neq h$, and that it is not possible to observe the start of the plant at the beginning of each period,[3] but it is possible to observe the quality of the water of the lake at the end of each round.

The net joint surplus (which represents the economic rent associated with a cleaner lake) is given by

$$(2) \ S = f(I) - \sum_j c^j (i^j) \qquad \forall j = (A, B)$$

where $I = (i^A, i^B)$ is the investment vector. Let us assume that the surplus sharing rule is designed according to the generalised Nash bargaining solution

$$(3) \ u^j = \frac{S}{2} \qquad \forall \ j = (A, B).$$

The optimal level of surplus is given by the unique investment vector which maximises the expression (2)

$$(4) \ I^* = arg \ max \ S$$

with I^* which satisfies the following first order conditions:[4]

$$(5) \ \begin{cases} f_A(I^*) = c^{,A} (i^A *) \\ f_B(I^*) = c^{,B} (i^B *) \end{cases}.$$

Each agent payoff associated with $I=I^*$ is given by $\pi^j *$, with $j=(A, B)$ and $S^*=S(I^*)$.

The cooperative solution ($\pi^A *$, $\pi^B *$), which represents the bilateral "commitment strategy" equilibrium, is not enforceable in the one-shot *meta*-game (*start the process, don't start it*) if each agent is induced to free-ride or hold-up.

In the case of free-riding, the payoff of firm A, for instance, is given by $\pi^A(F)$ [5] which is determined by the value if the investment which solves

---

[3] The model is developed on the basis of Putterman and Skillman (1992).

[4] We assume that $Q(I^*) \geq Q*$, where $Q*$ is the quality threshold for CPR sustainability.

[5] To simplify, we assume a positive investment level also in the case of free-riding, i.e. the process is started at its sub-optimal level.

$$(6) \quad i^A = argmax \left[\frac{f(i^A | I^*_{-A})}{2} - c(i^A) \right]$$

where I^*_{-A} is the investment vector characterised by the optimal investment choice of all agents with the exception of agent A (in the two-agent setting, it is the investment vector containing the optimal choice of firm B). For this a strategy to be the optimal response of agent A, we should have $\pi^A(F) > \pi^A *$.

In the case of hold-up the opportunistic agent is induced to realise the maximum joint surplus, with respect to the free-riding case. However, in order to increase his ex-post contractual power, the opportunistic agent is induced to ex-post renegotiate the surplus sharing rule. In this case the payoff of firm A will be given by

$$(7) \quad \pi^A(H) = \alpha S^* > \frac{S^*}{2}$$

where α is agent A's ex-post contractual power, with $a \in (1/2, 1]$.

For hold-up strategy to be the optimal response to B's commitment, we should thus have $\pi^A(H) > \pi^A *$. As we have noticed before in the case of hold-up strategy, agent A will delay any investment decision (to start the process) until he observes B's investments; once agent B is locked-in by the investment made, agent A will renegotiate the surplus sharing rule, and will selects the efficient investment which generates the joint surplus S^*, only after having observed the efficient choice made by the other agent.

Consider now the choice of the optimal response to the other firm's free-riding or hold-up, in the one-shot game. In this case, the payoff of firm A is given by $\pi^A{}^\circ$ generated by the following investment

$$(8) \quad i^A{}^\circ = argmax \frac{f(i^A | I^\circ_{-A})}{2} - c(i^A).$$

The game in the matrix below is the "one-shot" version of the two-period game shown above (the "reciprocity strategy" is represented by the diagonal through the top right elements).

Table 3: The one-shot non-cooperative game

			FIRM B		
			FREE-RIDING $i^B(F)$	HOLD-UP $i^B(H)$	COMMITMENT i^B*
FIRM A	FREE-RIDING	$i^A(F)$	$\pi^{A}\circ;\pi^{B}\circ$	$\pi^{A}\circ;\pi^{B}\circ$	$\pi^A(F);(\pi^B*-\phi);$
	HOLD-UP	$i^A(H)$	$\pi^{A}\circ;\pi^{B}\circ$	$\pi^{A}\circ;\pi^{B}\circ$	$\pi^A(H);(\pi^B*-\xi);$
	COMMITMENT	i^A*	$(\pi^A*-\phi);\pi^B(F)$	$(\pi^A*-\xi);\pi^B(H)$	$\pi^A*;\pi^B*$

Result 1 - Free-riding as a Nash equilibrium

In the game above if $\forall\ j=(A,\ B)$, ξ is such that $(\pi^j*-\xi)<\pi^{j}\circ$, with $\phi>\xi>0$ and $\pi^j(F)>\pi^j(H)>\pi^j*>\pi^{j}\circ>0$ then the non-cooperative game has a unique Nash equilibrium denoted as "bilateral underinvestment" $I°$.

Result 2 - Hold-up as a Nash equilibrium

In the game above if $\forall\ j=(A,\ B)$, ξ is such that $(\pi^j*-\xi)>\pi^{j}\circ$, ϕ is such that $(\pi^j*-\phi)<\pi^{j}\circ$ where $\phi>\xi>0$, and $\pi^j(F)<\pi^j(H)$ then the non-cooperative game has three Nash equilibria (two "unilateral hold-up strategies") and one bilateral free-riding strategy equilibrium.

Result 3 - Commitment as a Nash equilibrium

In the game of table 4 if $\forall\ j=(A,\ B)$, ϕ is such that $(\pi^j*-\phi)>\pi^{j}\circ$, where $\phi>\xi>0$ and $\pi^j(H)<\pi^j(F)<\pi^j*$ then the non-cooperative game has a unique Nash equilibrium associated with bilateral commitment strategies.

Depending on the value of ex-post monetary transfers between agents ϕ and ξ we have multiple Nash equilibria for the game described above. Results 1-3 have the property of rendering explicit the different conditions which characterise alternative forms of interdependence between agents and some possible "policy prescriptions" to enforce the efficient outcome.

Let us consider, for instance, Result 2. The assumption that ξ is such that $(\pi^j * - \xi) > \pi^{j\circ}$ may be written as follows: ξ is such that $(\pi^j * - \xi) > \pi^{j\circ}$ with $(\pi^j * - \xi) < \pi^{j\circ}$. In other terms, the opportunistic agent will always give to the other agent a payoff greater than his next best alternative, with ξ defined as follows

$$(9) \quad \xi = \alpha S * - \frac{S *}{2} = (\alpha - \frac{1}{2})S* = [\pi^A(H) - \pi^A *],$$

where α is agent A ex-post contractual power, with $\alpha \in (1/2, 1]$.

In order to avoid the hold-up equilibrium in Result 2, it is therefore sufficient to make a separate contract about the size of ξ. For example, each agent may decide to pay a *ex-ante* monetary bond equal to ξ, so that the expected joint surplus will be equal to $(S*-2\xi)$ and each agent will *ex-post* receive ξ, after having efficiently invested. Since, the opportunistic agent will thus receive (ξ), in the case of hold-up, each agent will thus be induced to efficiently invest because now the efficient investor will receive a payoff equal to $(\pi^j * - \xi) > \pi^{j\circ}$ in the case of other agent hold-up. This assumption is equivalent to having a "no ex-post renegotiation condition" on ξ.

However, the solution suggested above may be difficult to implement if free-riding yields a payoff greater than commitment strategy. In Result 2, for example, even in the case of the "no ex-post renegotiation condition" on ξ, the assumption $(\pi^j * - \phi) < \pi^{j\circ}$ implies the persistence of free-riding (a Nash equilibrium characterised by bilateral under-investment). In this a case the size of ϕ determines an under-investment decision. We conclude that even if it is possible to deter hold-up by introducing a "no ex-post renegotiation condition" on ξ in our one shot commons game, the economic loss inflicted to the victim in the case of free-riding is too great $[(\pi^j * - \pi^{j\circ}) > \phi]$ to be effective and credible. As a consequence, we will thus have a Nash equilibrium characterised by bilateral under-investment also when hold-up is prevented but free-riding is still a possible outcome.

A solution of separate contracting on ϕ, like the one shown above for ξ, so that the expected joint surplus is equal to $(S^*-2\phi)$, may not be enforceable, since in the free-riding case, free-rider is not expected to invest at all (this is therefore the main difference between free-riding and hold-up strategies).

Given (8), we conclude that bilateral free-riding and bilateral hold-up lead to the same under-investment configuration in the above one shot game.

However, under-investment is a dominant strategy as long as $\pi^j(F) - \pi^{j*} = \gamma > 0$ in the case of free-riding; whereas in the hold-up case, the opportunistic agent simply delays his investment in order to renegotiate the terms of trade.

In order to deter free-riding it is therefore necessary to resort to self-enforcing sanctioning strategies in a repeated game setting, as shown in the next section.

.

The repeated *commons game* with zero exit options: "trigger strategies" and "optimal penal codes"

Let us assume, now, that an outcome like the one described in Result 1 occurs in our setting. Suppose that the resulting free-riding is only the first stage of a repeated game. In a repeated game *ad infinitum*, agent j's actual payoffs are:

$$(9) \quad \Pi^j = \sum_{t=0}^{\infty} \delta_j \pi^j(u^{jt}, i^{jt}) \qquad \forall \; j = (A, B)$$

where δ is agent j's time discount factor, and $\delta<1$.

It is well known that the resulting "supergame" may lead to an efficient equilibrium, as long as agents adopt credible trigger strategies, and as long as players do not discount the future too heavily.

A straightforward problem in the implementation of trigger strategies is that punishment of free-riders, always also inflicts economic damage on the victims of free-riders. This situation is particularly evident in the case of environmental policy games, as in our example of the lake. The greater is the sanction, the higher the probability of an environmental disaster (which also leads to a breakdown of firms' production processes in the example of the lake).

The notion of "optimal penal code" formulated by Abreu (1988) may partially overcome the problem of "excessive punishment", suggesting a way of finding the most effective trigger strategies consistent with the players' preferences and game structure. The characteristic feature of an optimal penal code is to provide a sanction mechanism which works for a limited number of periods, say τ, after which the cooperative game can restart.

As Putterman and Skillman (1992) have pointed out, this is a stick-and-carrot mechanism, where "a penalised player must first take the medicine of unpleasant outcomes before being allowed to return to more desirable conditions". Through the penal code mechanism, each agent is induced to participate in his own punishment by the credible threat that the punishment will be reapplied if the agent deviates from cooperative behavior.[6] In other terms, there is a mixed strategy of cooperation, reciprocity and free-riding.

At the same time, the victim of free-riding only sustains the damage associated with the inefficient outcome for a limited number of rounds.

Consider the following deterrence strategy $Z(Q^*,Q^\circ)$[7] in our "lake game":

$$\begin{cases} Q^* = \text{cooperate until a deviation is verified, then switch to } Q^\circ \\ Q^\circ = \text{play Nash equilibrium } I = I^\circ \text{ for t periods then switch to } Q^*. \end{cases}$$

We obtain the following Result.

Result 4

Given the "lake game" in Result 1, with a unique under-investment Nash equilibrium $I=I^\circ$, and given the strategy $Z(Q^*, Q^\circ)$ defined as above, the one-shot defection from full cooperation (commitment strategy) is never profitable if :

$$(10) \quad \frac{\pi^{j\,*}}{1-\delta} \geq \pi^j(\bullet) + \frac{\delta}{1-\delta}\pi^j_{Q^\circ} \qquad \forall \; j = (A, B)$$

---

[6]   The threat of reversion to the punishment is credible because the non-cooperative strategy is an equilibrium in the one shot game.

[7]   Putterman and Skillman (1992).

where $\pi_{Q^\circ}^j$ is defined as

(11) $\quad \pi_{Q^\circ}^j = (1 - \delta^\tau)\pi^{j\circ} + \delta^\tau \pi^{j} *$ for some integer $\tau > 0$ to be determined.

In the one-shot game, $\pi^j(\bullet)$ may apply to free-riding or hold-up strategies. In the case of hold-up (11) may be written

(12) $\quad \dfrac{\pi^{j} *}{1 - \delta} \geq (\xi + \pi^{j}*) + \dfrac{\delta}{1 - \delta}(1 - \delta^\tau)\pi^{j\circ} + \delta^\tau \pi^{j} *$

$\forall\, j = (A, B)$.

From (12) we can argue that ξ, (i.e. the possibility and extent of ex-post renegotiation in the case of hold-up) will affect the probability of full cooperation. For instance, if ξ, is too high to prevent hold-up, a ex-ante side-payment equal to ξ may promote the efficient outcome.

Result 5

Given the expressions in Result 4, the value of τ determines the range of δ values which sustain the efficient cooperative outcome.

Proof: Rearranging (11) it follows that

(13) $\quad \delta \geq \dfrac{\pi^{j}(F) - \pi^{j} *}{\pi^{j}(F) - \pi_{Q^\circ}^j}$ in the case of free-riding $\forall\, j = (A,$

$B)$

(14) $\quad \delta \geq \dfrac{(\xi + \pi^{j}*) - \pi^{j} *}{(\xi + \pi^{j}*) - \pi_{Q^\circ}^j}$ in the case of hold-up $\forall\, j = (A, B)$

Since $\pi_{Q^\circ}^j = (1 - \delta^\tau)\pi^{j\circ} + \delta^\tau \pi^{j} *$, $\pi_{Q^\circ}^j$ is decreasing and strictly convex in τ. For τ near to zero, $\pi_{Q}^j \cong \pi^{j} *$ and $\delta \geq \dfrac{(\xi + \pi^{j}*) - \pi^{j} *}{(\xi + \pi^{j}*) - \pi^{j} *} \cong 1$

133

(in the case of hold-up),[8] while as long as τ grows, $\pi_Q^j \cong \pi^{j\circ}$ and

$$\delta \geq \frac{(\xi + \pi^{j}*) - \pi^{j}*}{(\xi + \pi^{j}*) - \pi^{j\circ}}$$ which implies a widening of the range of the value

of the discount factor which is compatible with full cooperation, given that

$$\frac{(\xi + \pi^{j}*) - \pi^{j}*}{(\xi + \pi^{j}*) - \pi^{j\circ}} < \frac{(\xi + \pi^{j}*) - \pi^{j}*}{(\xi + \pi^{j}*) - \pi_{Q\circ}^{j}} \cong 1.$$ The larger τ, the more

effective the threatened punishment, and the wider the range value which allows full cooperation.

When (10) holds, then a strategy like $Z(Q^*,Q^\circ)$ is credible, if I° is a Nash equilibrium in the one-shot game. Paradoxically, the one-shot "tragedy" is an optimal enforcement device, under the optimal penal code hypothesis.

However, as we have already noticed, the greater the number of rounds in which Nash equilibrium I° is reached, the longer the economic loss for the victim of opportunistic behaviour.

Result 6

Depending on the value of ξ and γ, if

$$(15) \quad \frac{\pi^{j}* - \phi}{(1 - \delta)} \geq (1 - \delta^{\tau})\pi^{j\circ} + \delta^{\tau}\pi^{j}* \qquad \text{in the case of free-riding}$$

$$\forall \; j = (A, B)$$

$$(16) \quad \frac{\pi^{j}* - \xi}{(1 - \delta)} \geq (1 - \delta^{\tau})\pi^{j\circ} + \delta^{\tau}\pi^{j}* \qquad \text{in the case of hold-up}$$

$$\forall \; j = (A, B)$$

then the victim of opportunistic behaviour may prefer to make a side-payment to the opportunistic agent rather than adopt the sub-optimal choice for very many rounds.

---

8   The case of free-riding brings to the same results.

Conclusion 1

For a given time discount factor δ, it is possible to sustain full cooperation equilibrium by adoption of an optimal penal code. However, the determination of τ, in the penal code, is subject to the expected economic loss of the victim of free-riding or hold-up. Agents may then strategically determine the value of (τ, ξ, γ) to reach full cooperation (where γ is the sum to be transferred to the free-rider, $\pi^j(F) - \pi^{j*} = \gamma > 0$).[9]

Results 4-5 assume that each agent can't avoid the punishment when acts as an opportunist, or the loss inflicted by opportunistic behaviours when he/she is the victim. In other terms, we have implicitly assumed that "exit" the relationship is not an available strategy. The following section removes such an assumption in order to evaluate how the incentives to full cooperation may be affected when agents may simply "exit" the relationship.

The *commons game* with exogenous exit options: "escape" *versus* "lock-in" equilibria

Let us assume that an exit option is available for each agent in the previous "lake game", and that this options represents the highest payoff obtainable outside the game. In other words, agents may now coordinate their actions with new agents (for example, agent A may start a joint-venture with another firm, say C, which sells clean water, or agent A may move to another lake, and so on).[10] Let $V^j > 0$ be agent j's exit option $\forall\ j = (A, B)$, measured in monetary terms. In the repeated game of section 4, this may modify some of the conclusions.[11]

Let us again consider (10) and assume that the exit option of the potential opportunist is

$$(18) \qquad V^j > \pi^j_{Q^\circ}. \quad \forall\ j = (A, B).$$

---

[9] Thus, in the case of free-riding, the (15) becomes: $\dfrac{\pi^{j*} - \gamma}{(1-\delta)} \geq (1-\delta^\tau)\pi^{j\circ} + \delta^\tau \pi^{j*}$.

[10] In a very interesting article, Lin (1990) surveys alternative exit options in Chinese agricultural co-operatives, showing a trade-off, in terms of efficiency, between strategic flexibility inside the cooperative and external mobility in the market.

[11] We assume that the choice of 'exit' terminates the game.

When a trigger strategy like $Z(Q^*, Q^\circ)$ is adopted, the opportunistic agent may select the exit option and "escape" in order to avoid the punishment, as in the following result.

Result 7

Given Result 4 and positive exit options V^j, when a trigger strategy like $Z(Q^*, Q^\circ)$ is implemented, full cooperation may be sustained by an optimal penal code if there is a τ for which

$$(19) \quad \delta^V \geq \frac{\pi^j(F) - \pi^{j*}}{\pi^j(F) - \pi^j_{Q^\circ}(\tau)} > \frac{\pi^j(F) - \pi^{j*}}{\pi^j(F) - V^j}$$

in the case of free-riding, $\forall \ j = (A, B)$;

$$(20) \quad \delta^V \geq \frac{(\xi + \pi^{j*}) - \pi^{j*}}{(\xi + \pi^{j*}) - \pi^j_{Q^\circ}(\tau)} > \frac{(\xi + \pi^{j*}) - \pi^{j*}}{(\xi + \pi^{j*}) - V^j}$$

in the case of hold-up, $\forall \ j = (A, B)$.

Proof. We can easily see that if $V^j > \pi^j_{Q^\circ}$, the range of values of the discount factor which sustains the cooperation is now reduced and punishment is ineffective.

Conclusion 2

When positive exit options are introduced in the commons game, the efficient amount of τ will be determined by the value of exit options which then act as a constraint. In such a setting, agents face a trade-off between the effectiveness of punishment strategies, which calls for a high value of τ and a low value of $\pi^j_{Q^\circ}$; and the exit options constraint which requires a low value of τ (a high value of $\pi^j_{Q^\circ}$), so that $\pi^j_{Q^\circ} > V^j$ always holds.

Result 8

Given V^j, side-payments to prevent free-riding and/or hold-up are subject to the value of exit options, which therefore acts as a constraint for the determination of side-payments (MacLeod and Malcomson, 1993; Osborne and Rubinstein, 1990), so that

(17) $\gamma = \gamma(V^J), \xi = \xi(V^J) \quad \forall \ j = (A, B).$

Proof: When $V^j > \pi^j_{Q^\circ}$, "free-riding" is a dominant strategy if

(18) $\pi^j(F) \geq \dfrac{\pi^{j\,*}}{1-\delta} - \dfrac{\delta}{1-\delta} V^j \qquad \forall \ j = (A, B)$

In order to prevent free-riding it is therefore necessary to make the potential free-rider a side-payment at least equal to

(19) $\gamma = [\pi^j(F) - \dfrac{1}{1-\delta}(\pi^{j\,*} - \delta V^j)] \qquad \forall \ j = (A, B);$

"Hold-up" is a dominant strategy if

(20) $\xi \geq \dfrac{\delta}{1-\delta}(\pi^{j\,*} - V^j) \qquad \forall \ j = (A, B)$

In order to prevent hold-up it is therefore necessary to make the potential opportunist a side-payment at least equal to $\xi = \dfrac{\delta}{1-\delta}(\pi^{j\,*} - V^j).$

Conclusion 1 is therefore modified as follows.

Conclusion 3

Given the time discount factor δ and exit options V^j, $\forall \ j = (A, B)$, full cooperation equilibrium can be sustained by adoption of an optimal penal code. However, the determination of τ in the penal code is subject to the expected loss of the victim of free-riding or hold-up. Agents may then strategically determine the value of $[\tau, \gamma = \gamma(V^J), \xi = \xi(V^J)]$ to achieve the full cooperation.

Conclusions 2 and 3 show that the rules of the commons game are modified by the introduction of exit options. An important feature of the above proofs is that they may be adopted also for the agents involved in the game. The following results sum up the conditions for sanctioning free-riding or hold-up, depending on the distributions of exit options between agents.

Result 9 – Effective Sanction: exit options as "lock-in"

Given the time discount factor δ, and agents' exit options $V^j \; \forall \; j = (A, B)$, if in the one-shot game we have that

$$
\begin{cases}
(21) \quad \delta \geq \dfrac{\pi^i(\bullet) - \pi^j *}{\pi^i(\bullet) - \pi_{Q°}^j} \\[2ex]
(22) \quad \pi_{Q°}^j > V^j \qquad\qquad \forall \; j,h = (A,B) \\[2ex]
(23) \quad \pi_{Q°}^h > V^h
\end{cases}
$$

thus a trigger strategy like $Z(Q^*, Q°)$ will enforce an effective sanction against agents' opportunism $[\pi^j(\bullet)$ refers to free-riding or hold-up strategies], given that exit options are never profitable for both the agents involved in the game (*agents lock-in*).

Result 10 – Ineffective Sanction: exit options as "escape"

If equations (22) and (23) are not satisfied in result 9, and exit options are always profitable for both agents involved in the game (*agents escape*) then a trigger strategy like $Z(Q^*, Q°)$ will be ineffective.

Conclusion 4

Under the conditions expressed in Result 9, agents may define an optimal side-payment (such that $\pi_{Q°}^j + (side\text{-}payment) \geq V^j$) and an optimal level of τ, such that full cooperation occurs.

Result 11 – Effective sanction: lock-in for the opportunist

Let agent j be the opportunistic agent, following free-riding or hold-up strategies, and h be the committed agent. If equation (23) is satisfied but not equation (22), then agent h may sanction the potential opportunist simply by exiting, rather than adopting a trigger strategy like $Z(Q^*,Q^\circ)$, since agent h's exit options are more profitable than $\pi_{Q^\circ}^j$, while that of agent j the exit option is never profitable. In this case the "threat" of exit is credible and the opportunist is induced to efficiently invest.

Result 12 – Ineffective sanction: committed agent lock-in (hold-up)

Let agent j be the opportunist (free-riding or hold-up) and h be the committed agent. If equation (22) is satisfied but not equation (23), then agent j may threaten to exit unless trading terms are renegotiated, since for agent j's exit options are more profitable than $\pi_{Q^\circ}^j$, while that of agent h is never profitable. Hence, the cooperative solution will only be reached if the committed agent h makes a side-payment to the opportunist j, such that $\pi_{Q^\circ}^j + (side\text{ - }payment) \geq V^j$. Otherwise, bilateral under-investment will occur.

Result 13 – Bilateral "escape" as an enforcement device

If equations (22) and (23) are not satisfied, a trigger strategy like $Z(Q^*,Q^\circ)$ is never effective. However if equation (24) holds

$$
\left\{
\begin{array}{ll}
(21) & \delta \geq \dfrac{\pi^i(\bullet) - \pi^{j}*}{\pi^i(\bullet) - \pi_{Q^\circ}^j} \\[4mm]
(22) & \pi_{Q^\circ}^j < V^j \\[2mm]
(23) & \pi_{Q^\circ}^h < V^h \\[2mm]
(24) & S* > V^j + V^h \ \ (super\ additivity)
\end{array}
\right.
\qquad \forall\ j,h = (A,B)
$$

then the only possible sanction is "escape", which is not a credible threat because exit will destroy the expected quasi-rents.

Results 6-13 show how the rules of the games are modified by the introduction of exit options. In this context, side-payments may play a strategic role in deterring free-riding and hold-up. According to this view, a "mirror game" can be designed as a linked game which determines agents' payoffs subject to the optimal value of side-payments required in the "principal game". Examples could be transfer of technology or cooperation in development policies, commercial exchanges and so on.

However, as we have stressed, in order to deter opportunistic behaviour, especially hold-up, it is also necessary to "link" the "timing" of side-payments. If a side-payment is received before cooperation, then hold-up may occur, leading to renegotiation of trading terms. Thus introducing a linked game may even enhance the incentive to select opportunistic behaviour rather than sustain full cooperation.

Unfortunately, in most cases, it may only be possible to time side-payments at very high monitoring costs. These costs may be extremely high when exogenous shocks change payoffs over time. In particular, when side-payments are subject to exit options, exogenous changes in exit options may render a previous side-payment insufficient and contracting schemes on side-payments $\gamma = \gamma(V^J), \xi = \xi(V^J)$ will result in an incomplete contract, due to uncertainty on the future realisation of V^J (MacLeod and Malcomson, 1993).

In this perspective, conclusions 2-4 show how essential it is to take into account parties' exit options in order to determine optimal penal codes and side-payments to sustain full cooperation.

Thus, a linked game which simply determines the value of side-payments without explicitly valuing taking into account the expected value of parties' exit options, may result in an ineffective enforcement device and it may even encourage opportunistic behaviours. This conclusion calls for an analysis of the strategic role of exit options to enforce full cooperation outcome.

The *commons game* with endogenous exit options: the strategic role of asset specificity in sustaining full cooperation

In this section, we no longer assume exogenous exit options. We assume that agents may build a "linked game" characterised by a Nash equilibrium which ensures payoffs equal to the optimal value of transfers ξ and γ, so that full cooperation, in the "lake game", is achieved.

From Result 8 we know that when exit options are binding the optimal value of side-payments in the case of opportunistic behaviour is determined by the value of agents exit options, so that $\gamma = \gamma(V^J), \xi = \xi(V^J)$. We therefore assume that it is possible to build a linked game characterised by a positive vector of investments, denoted by $K = (k^A, k^B)$ such that

$$\frac{dV^j(K)}{dK} < 0, \forall \ j = (A, B).$$ In other words, the linked game, ones started,

reduces the exit options. We can imagine, for instance, that adoption of a technology will endogenously enhance the bilateral dependency of agents

A and B in terms of the degree of asset specificity developed. Investments $K = (k^A, k^B)$ may thus be viewed as *specific* investments which generate the required side-payments, but at the same time increase agents' quasi-rents, decreasing their best alternative opportunity outside the relationship. Specific investments provide a higher ex-post value with respect to the ex-ante parties' outside options only if the underlying transaction takes place. Once made, a specific investment will lock the investors into the contractual relationship by raising their ex-post exit costs: outside the transaction, the ex-post value of specific assets or investments will thus be lower than their best ex-ante outside options. Agents who make specific investments are then vulnerable to counterpart's post-contractual opportunism. Thus we have the following Result.

Result 14 – Asset co-specificity as an enforcement device

Given Result 13, if it is possible for the agents involved in the game to select a vector investment $K = (k^A, k^B)$ such that $\frac{dV^j(K)}{dK} < 0, \forall j = (A, B)$, and if it is possible to select a value of $K=K*$ such that[12]

$$\begin{cases} (25) & \pi_{Q^\circ}^j \geq V^j(K*) \\ (26) & \pi_{Q^\circ}^h \geq V^h(K*) \qquad\qquad \forall j, h = (A, B) \\ (27) & S*-C(K*) > V^j(K*) + V^h(K*) \end{cases}$$

then full cooperation may be achieved.

Proof   In this case, full cooperation may be reached at positive enforcement costs $C(K*)$, since the range of values of the discount factor that sustains full cooperation is increased and the punishment introduced by optimal penal codes is effective, whereas exit would result in complete dissipation of the expected quasi-rents for both agents.

Result 14 shows how a linked game should be structured in order to enforce full cooperation. In other words, the linked game (aid policy,

---

[12] Where $C(K)$ is the cost function associated with the investment vector K, with $c'(K) > 0, c'' > 0, \forall k^j \geq 0$.

technological cooperation, and so on) should not only transfer the appropriate level of side-payments, but should also increase agents' co-specificity, increasing their exit costs, so that exit will penalise opportunistic behaviour. Once the exit option is not longer available, agents may also recur to traditional penal codes to prevent opportunistic behaviour.

Notice that the cost associated with the specific investment vector K can be explained as the cost of the endogenous enforcement that agents have to sustain to reach the full cooperation equilibrium. Such enforcement costs should be compared with the cost of monitoring agents' behaviour at each round.

In Result 14 we assumed a very generic exit option function, assuming the existence of a vector K^*, which affects both parties' exit options, in the same way. However, strategic interdependence may also be affected by the fact that exit options may vary in an asymmetric way, so that the degree of specificity varies between agents. In some cases, a solution like the one assumed in Result 12 may therefore also occur with endogenous exit options.

Conclusions

Departing from the traditional problem of free-riding in CPRs, we have analysed free-riding and hold-up as alternative forms of opportunistic behaviour, such as free-riding and hold-up, which may co-exist in some setting. While free-riding may when under-investment is a dominant strategy in the standard PD game, hold-up is possible in cases of coordination failure, as in the "chicken game", when each agent prefers to delegate implementation of the cooperative solution to the other agent to. In this setting, in the absence of coercion or binding agreements, each agent may "wait" to cooperate until the counterpart is committed to cooperate and then ask for renegotiation of contractual terms.

The unified framework proposed here, determines the optimal value of side-payments to be transferred in the case of free-riding and hold-up, thus distinguishing games structured as PD from CG.

However, when positive exit options are available to the parties, depending on the value of exit options, optimal penal codes (Abreu, 1988) may be ineffective and simple side-payments – as in the "linked game" approach - may not be the best strategy to achieve full cooperation, unless it is possible to optimise the timing of the side-payments transfers and the actions selected by the players. This timing might however require very high costs and thus simple side-payments may be ineffective for enforcing full cooperation.

We suggested than that an optimal linked game should be structured so as to enhance agents' exit costs, by the introduction of asset specificity or specific investments which affect parties' exit options. Under given conditions, we showed that full cooperation may be achieved, at positive enforcement costs $C(K^*)$, when the choice of exit options would result in complete dissipation of the expected quasi-rents of both agents. We stressed that in order to enforce a bilateral incomplete contract, the agents involved could lock-in their counterparts, reducing their exit options. To enforce contractual performance, parties involved in an incomplete contract could also assets specificity as an endogenous enforcement device. Analysis of the enforcement role of endogenous variations of exit options may therefore suggest new policy prescriptions based on the development of specific investments (delayed side-payments or technology transfers which require future technological cooperation) to induce parties to cooperate in a long-term perspective. However, the enforcement role of exit options depends mainly on asymmetries between the parties involved in the contractual relationship, so that exit options may enhance or destroy cooperation depending on the degree of contractual dependency of the potential opportunist.

REFERENCES

Bromley D.W., (1989) *Economic Interest and Institutions*, Blackwell.

Bromley D.W., (1991) *Economics and Environment*, Blackwell,.

Carraro C. and Siniscalco D., (1991a), "Strategies for international protection of the environment", *CEPR Discussion Paper* n. 568, 8.

Carraro C. and Siniscalco D., (1991b) Environmental Innovation Policy and International Competition, CEPR, discussion paper n. 525.

Carraro C. and Siniscalco D., (1992) The International Dimension of Environmental Policy, in *European Economic Review*, n. 36, pp. 379-387.

Chichilniski G., Heal G., Pagano U., (1994), "Property Rights and Returns to Scale: Patents, Firm, Market Failure", *Quaderni del Dipartimento di Economia Politica*, n.161, University of Siena.

Chung, T.Y., (1996), "On Strategic Commitment versus Investment", *American Economic Review, Papers and Proocedings*, pp. 437- 441.

Coase, R., (1960), "The Problem of Social Costs", *Journal of Law and Economics*, 3.

Cornes R., Sandler T., (1983) On Commons and Tragedies, *American Economic Review*, nø 73, 1983, pp. 787-792.

Cornes R., Sandler T., (1985), Externalities, Expectations and Pigouvian Taxes, *Journal of Environmental Economics and Management*, n. 12, pp. 1-13.

Cornes R., Sandler T., (1986), *The Theory of Exernalities, Public Goods and Club Goods*, Cambridge Un. Press, Cambridge.

Dasgupta P., (1990) The Environment as a Commodity, in *Oxford Review of Economic Policy*, vol. 6, n. 1.

Edlin A., Reichelstein S., (1996), "Holdups, Standard Breach Remedies and Optimal Investment", *American Economic Review*, Vol. 86, n. 3, pp. 479-501.

Franzini, M. (1995) "Post-contractual opportunism and the market", Mimeo, University of Siena.

Franzini, M. (1999) "Adaptation and Opportunism in Political and Economic Markets", in S. Bowles, M. Franzini, U. Pagano (Eds) *The Politics and Economics of Power*, Routledge.

Franzini M. and A. Nicita (Eds), (1999) *Economic Institutions and Environmental Policy*, forthcoming, Ashgate.

Grossman S. and Hart O. (1986) "The Costs and Benefits of Ownership: A Theory of Vertical and Lateral Intergation", *Journal of Political Economy*, 94:691-719.

Guttmann J.M., (1978) "Understanding Collective Action: Matching Behavior", *American Economic Review*, n. 68, pp. 251-255.

Hardin, G. (1968) "The Tragedy of the Commons", *Science*, 13/12.

Hart, O., (1995), *Firms, Contracts and Financial Structure*, Oxford University Press, Oxford.

Hart, O. and Moore, J., (1990), "Property rights and the Nature of the Firm", *Journal of Political Economy*, 98, pp. 1119-1158.

Leitzel, J., (1993), "Contracting in Strategic Situations", *Journal of Economic Behavior and Organization*, 20, pp. 63-78.

Lin, J. Y. (1990), "Collectivizations and China's Agricultural Crisis in 1959-1961", *Journal of Political Economy* 98, 6:1228-1252.

MacLeod, B. (1993), "The Role of Exit Costs in the Theory of Cooperative Teams: A Theoretical Perspective", *Journal of Comparative Economics*, 17521-529.

MacLeod, B. and Malcomson, J., (1993), "Investments, Hold-up, and the Form of Market Contracts", *American Economic Review*, 83, pp. 811-837.

Magrath W., (1989) "The Challenge of the Commons: the Allocation of Nonexclusive Resources", *The World Bank, Environment Department*, WP n. 14.

Nicita A. (1997), *Contratti Incompleti, Investimenti Specifici ed Enforcement endogeno*, tesi di Dottorato, Universita di Siena.

Nicita A. (2001), "The Firm as an Evolutionary Enforcement Device", in A. Nicita and U. Pagano (Eds) *The Evolution of Economic Diversity*, Routledge.

Olson M., (1965), *The Logic of Collective Action. Public Goods and the Theory of Groups*. Harvard University Press.

Ophuls, W. (1973) "Leviathan or Oblivion" in H.E. Daly (Ed.) *Toward a Steady State Economy*, Freeman.

Osborne M.J., Rubinstein A. (1990), *Bargaining and Markets*, Academic Press,

San Diego CA.

Ostrom, E. (1990), *Governing the Commons*, Cambridge University Press.

Pagano, U. (1999) "Is Power an Economic Good? Notes on Social Scarcity and the Economics of Positional Goods" in S. Bowles, M. Franzini, U. Pagano (Eds) *The Politics and Economics of Power*, Routledge.

Putterman, L. and Skillman, G., (1992), "The Role of Exit Costs in the Theory of Cooperative Teams", *Journal of Comparative Economics*, 16, pp. 596-618.

Salop, S.C., Scheffman D.T., (1983) "Raising Rivals" Costs', *American Economic Review* 73, pp. 267-71

Schelling, T., (1960), *The Strategy of Conflict*, Harvard University Press,Cambridge, Ma.

Tirole, J., (1994), "Incomplete contracts: where do we stand?", mimeo, IDEI, Toulouse.

Williamson O., (1985), *The Economic Institutions of Capitalism. Firms, markets, Relational Contracting*, The Free Press, New York.

6. Facing Environmental "Bads": Alternative Property Rights Regimes for Local and Transnational Commons

Massimo Di Matteo and Antonio Nicita

Introduction: the transnational dimension of environmental externalities

Free-riding occurs in situations in which different agents interact in a common local or global environment, for example stateless societies, cases in which authorities fail to monitor and curb deviating behaviours and cases in which the parties fail to determine all the contingencies of an agreement or sign contracts that can be verified by a third party. In all these cases, opportunistic behaviour is encouraged by the low probability of sanctions.

Global transnational resources such as the atmosphere, oceans, stratosphere, tropical forests and so forth, are open access resources (resnullius), over which no single nation has exclusive sovereignty.

Transnational resources must therefore be safeguarded by voluntary agreements between states or through new international regulations and institutions.

In this context, the notion of externality changes and the discrepancy between social and private cost becomes a divergence between national and international cost (spatial extension) and between present and future cost (temporal extension). Conflict and coordination between agents interacting in the management and exploitation of environmental resources therefore regards relations between states, the objectives of which become maximisation of individual profit with minimisation of global damage.

In the case of global commons, there are externalities with respect to which countries are both the source and victim of global damage. For example, major transnational efforts are being made for global warming, caused principally by carbon dioxide emissions.

Because of the absence of supranational institutions with a capacity to dictate behaviour, impose sanctions and apply incentives, transnational externalities create a context in which the only possible way of safeguarding supranational environmental goods is for countries to use economic instruments and coordinate their policies. Here we review and discuss the effectiveness of various instruments.

Free trade of resources and movement of factors in a North-South context

It is worth recalling some aspects of economic development since the second world war to counter the idea that peoples of the so-called South are much more responsible for environmental degradation that those of the North.

This is untrue because the consumption of natural resources is directly proportional to per capita income which is much higher in the North. We therefore sustain that the economies of industrialised countries and the way in which they are run are the prime mover of environmental degradation.

They are extremely resource-intensive, encouraged by the low price of natural resources, especially energy. The high growth rate, at least in the first 25 years after the war, ran parallel to the extension of trade between North and South, with the South (e.g. Latin America, Africa) exporting basic goods (primary materials, natural resources, farm produce) and the North exporting manufactured goods. The extraction of resources *per se* causes severe environmental damage; examples range from oil extraction to exploitation of salt pans. The world market has mediated consumption and imports of natural resources by the North and production and exports by the South.

This situation is normally seen as the result of free trade and thus as efficient, in line with the traditional view that the specialisation of the South is due to the relative abundance of "natural resources" and that of the North due to the relative abundance of capital or specialised labour. Prices, especially those of natural resources, are therefore seen as a correct measure of social costs. The movement of factors (labour and capital) are in turn seen as effects of trade restrictions and hence as phenomena that restore equilibrium in the dynamics of the world economy.

This view will be criticised on the basis of Chichilnisky (1994) and Chichilnisky and Di Matteo (1997) and the inefficiency of this situation will be demonstrated. The existing specialisation may be the result of institutional factors, specifically, the particular distribution of property rights of natural resources. It can be argued that the North consumes and the South produces an excess of resources with respect to a Pareto-efficient situation. In other words, the price of resources is lower than it would be for Pareto efficiency.

Pareto efficiency of free trade

According to traditional international trade theory based on the Heckscher-Ohlin (HO) approach, free trade leads to achievement of Pareto efficiency (for detailed analysis see eg. Gandolfo, 1998). The demonstration relies on appropriate assumptions, such as the existence of a community indifference curve that coherently represents the preferences of all citizens of a country.

It therefore follows that equilibrium achieved by opening international trade enables a country to reach a higher (i.e. Northeast shifted) community indifference curve than could be attained in autarchy. International trade makes this possible because it frees countries from the need to produce the two goods in exactly the proportion dictated by internal demand, so that they can concentrate production in the more efficient sector.

As mentioned above, this approach has also been applied to the case in which international trade includes natural-resource-intensive goods, because they are not regarded as posing any particular problem for the application of HO theory.

Implications of different definitions of property rights

It is worth rendering explicit an assumption implicit to this line of thought.

A precise distribution of property rights on goods and resources underlies the efficiency achieved in a competitive market. It is assumed that economic agents have the right of private property to these goods and resources. However, there are other distributions of property rights besides private property, for example common property. In some such cases the market may not achieve a situation of efficiency.

To fully understand this point, it is as well to recall the concept of common property (for an interesting analysis see for example Bromley, 1992). A distinction exists between closed and free-access common property. Although the former is not private property, for a series of circumstances the community which has rights to the resource aims to conserve it and to exploit it in a lasting way. This occurs when a community residing stably near a resource such as a lake or forest, takes a long-term view in its economic decisions, including not only the present, but also future generations. The resource is not over-exploited because it has to remain for communities of the future. This generally is implied by the observation of collectively formulated rules.

Free-access common property is a different case. It generally comes about by a series of external events, such as the existence of alternative job possibilities which creates a different attitude towards extraction of the resource. In this situation the rules for conservation of the resource in time are no longer observed and behaviour is dictated by personal interests. The possibility of a salary from an alternative occupation (e.g. in a factory in the city), leads community members to seek similar value in their resource extraction activity. The rules of the community are no longer observed if they do not ensure an income similar to external alternatives. This becomes a reason for not considering, for example, a lake of common property, as the only source of subsistence for the future. In such a situation, closed common property becomes free-access; community members feel justified in extracting the resource up to a point in which their gains equal the salary that they could earn elsewhere. Quite similar results occur when there is no enforcement of private property rights.

It is well known (Dasgupta-Heal 1979) that in a situation of free-access common property, the quantity of work done by a worker is greater than the efficiency level which could be achieved in a different institutional milieu. It is therefore necessary to see whether natural resources are in common property regimes in the South. As far as they are, the model of Chichilnisky (1994), which we summarise below, formally demonstrates that the present situation of the world economy is not Pareto-efficient. We recall that in the South there are examples similar to free-access common property regimes: forests in Thailand, Nepal, the Amazon basin, pastures in Botswana, etc. (e.g. Pearce, Barber and Markandya 1990)

The Chichilnisky North-South heuristic model

Let us consider two regions or groups of countries: the North (industrialised countries) and the South (developing countries), two goods A and B and two factors. To produce the goods, labour L and a natural resource E are required. In the production of both goods, constant returns to scale and fixed production coefficients are assumed. The quantities of labour L and natural resource E are not constant but depend on the respective remunerations.

Even with fixed production coefficients, there is therefore substitutability between economic factors; when the relative prices of goods vary, so does the composition of production between A (labour-intensive) and B (natural-resource-intensive). Thus the aggregate ratio E/L varies because the two goods have different factor-intensities.

Competition in the good market and constant returns to scale imply zero extra profits in equilibrium such that:

$$P_B = a_1 P_E + c_1 w$$
$$P_A = a_2 P_E + c_2 w$$

where $P_A(P_B)$ is the price of good A(B), $P_E(w)$ the price of the natural resource (labour), $a_i(c_i)$ $(i=1,2)$ the coefficients of the natural resource (labour) in industries 1 (B) and 2 (A) respectively. The hypothesis that B uses the natural resource more intensely than A gives a positive value of $D = a_1 c_2 - a_2 c_1$.

In a general equilibrium model of this type, we have to assume a good as a numeraire, e.g. A, so that $P_A = 1$.

The prices of the two factors as a function of P_B, which now indicates the relative price of the goods, follow immediately:

$$P_E = \frac{c_2 P_B - c_1}{D}$$
$$w = \frac{a_1 - a_2 P_B}{D}$$

and hence:

$$\frac{\partial (P_E / P_B)}{\partial P_B} = \frac{c_1 P_B^{-2}}{D} > 0$$
$$\frac{\partial (w / P_B)}{\partial P_B} = \frac{-a_1 P_B^{-2}}{D} < 0$$

As we said, the supply of labour depends positively on real wage, w/P_B, as follows:

$$L^S = \beta w / P_B + L_0$$

where β and L_0 are greater than zero.

The supply of natural resource is also an increasing function of the price of the resource. The exact nature of this curve depends, however, on the distribution of property rights. It is assumed that a predominantly free-access common property rights regime exists in the South for extraction of the resource. What we said above about the lake gives a curve that for any level of q (the opportunity cost of labour), increases with increasing P_E. If there were private property in the South, the curve for extraction of the resource would be steeper because q/P_E would be equal to the marginal "private" product rather than to the marginal "common" product (where for private we mean "under conditions of private property" and for common, "under conditions of free-access common property"). If for simplicity we take this relationship to be linear, we can write:

$$E^S = \alpha P_E / q + E_0$$

where α and E_0 are greater than zero.

The parameter α is very large when there is common property of the resource (i.e. in the South) because it reflects the greater reactivity of supply of E to its price, with respect to the private property case. In this model, a large value of a expresses the so-called "tragedy of commons" which as we know leads to greater exploitation of the resource than under a private property regime.

The hypothesis is shown in the figure below, where E_{PP}^S is the supply of E under a private property regime and E_{CP}^S the supply of E in a free-access common property regime.

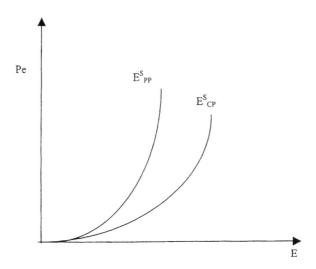

Figure 1: Supply for natural resource as a function of property regime

Demand for natural resource E^D and labour L^D depend on the quantities of goods produced:

$$E^D = a_1 B^S + a_2 A^S$$

$$L^D = c_1 B^S + c_2 A^S$$

where $B^S (A^S)$ is the supply of B (A).

Since markets for factors are in equilibrium we have:

$$L^D = L^S$$
$$E^D = E^S$$

In this model, with technology and preferences the same in the two countries (as required by the HO model), the difference in the property rights regime of the natural resource clearly leads to a difference in the comparative advantage of the two countries and hence to the possibility of an advantageous international trade. In other words, good B will be produced relatively more in the South than in the North and the South will therefore export it, with the effects mentioned in 2.1.

Exports of the South, X_B^S, are equal to the difference between supply and domestic demand, B^D:

$$X_B^S = B^S - B^D$$

whereas imports of A, X_A^D, are equal to the difference between demand, A^D and domestic supply:

$$X_A^D = A^D - A^S$$

It is assumed that in equilibrium, the value of exports equals that of imports *(equilibrium in the balance of trade):*

$$P_B X_B^S = X_A^D$$

The economy of the North

The North is characterised by an identical set of equations except for a different (lower) value of α that reflects the fact that the natural resource is in a private property regime in the North. Then there are the conditions of equality of prices of the two goods on the international market and those of the equality between supply of exports and demand for imports of each good. To close the model it is necessary to make some assumptions about the demand for one of the two goods in each country (the demand for the other good is of course implied by Walras' law which holds true in each economy); following Chichilnisky (1994), we consider the simple formulation below, for North and South:

$$A^D = A_0^D$$

Analytical solution of the model

It can be shown that the model reduces, through a series of substitutions, starting from the condition of equilibrium of good A on the world market,

$$A_0^D + A_0^D(N) = A^S + A^S(N)$$

to a second degree equation in a single unknown P_B :

$$[A(N)]P_B^2 + [A_0^D + A_0^D(N) + C(S) + C(N)]P_B - [V(S) + V(N)] = 0$$

where

$$A(N) = \alpha_N(c_1 c_2) / D^2$$
$$C(S) = (1/D)[c_1 E_0 - a_1 L_0 + (a_1 a_2 \beta + c_1 c_2 \alpha)/D]$$
$$C(N) = (1/D)[c_1 E_0 - a_1 L_0 + (a_1 a_2 \beta + c_1^2 \alpha_N)/D]$$
$$V(S) = \beta a_1^2 / D^2 + ac_1^2 / D^2]$$
$$V(N) = \beta(a_1^2)/D^2.$$

This equation has only one economically significant solution. The values of the other endogenous variables can be calculated by successive substitutions and comparative static analysis can be carried out simply using the theorem of the implicit function.

Main results of the model

The reasoning behind the above model can be summarised intuitively as follows. The two countries are equal except that the North has a private property regime on the resource whereas the South has a free-access common property regime. Efficiency at world scale is when both countries have a private property regime because in this case prices reflect marginal Pareto conditions. In the above case, this is not true for the South which has a greater supply of the resource than it would have under a private property regime. This has three main consequences. The first is that the extra supply of the resource in the South translates into excess supply of the good that uses the resource intensively and hence into a lower domestic relative price. When trade opens, the South finds itself exporting B which it produces more economically than the North, since preferences are unchanged and equal in both countries. The second is that the international price of the resource is different, namely less than in the efficient situation. When trade opens, in fact, the world price establishes at a level that ensures equality of world supply and demand of good *B*, which turns out to be intermediate between those prevailing independently in the two countries and is therefore always less than the price in the North (which is in the efficient situation). In fact the relative price of B in international equilibrium (terms of trade) is higher than the price prevailing in the South (in autarchy) but less than that prevailing in the North (in autarchy).

The third consequence is that the prices of the resources E and L become the same as a result of international trade (theorem of factor price equalisation). The configuration of prices and quantities therefore depends on the distribution of property rights on the resource, all other things being equal.

Other results

Other interesting results can be derived from this model with appropriate modifications. We can drop the assumption of equal technology in the North and South (typical of the HO model) and more realistically assume that the production coefficients in industries A and B are different in the North and South. However we continue to consider a situation in which the South exports a good that uses resource E intensively. As we know, when technology differs between countries, the factor price equalization theorem no longer holds.

Specifically, salaries will differ between North and South which will mean that labour will migrate from South to North, drawn by the wage difference. What are the consequences? We begin to see that an increase in L in the North and a decrease in L in the South will have overall contrasting effects on the relative production of the two goods and their prices.

To meet demand we therefore need to introduce an additional assumption, namely "technological dualism" (Chichilnisky & Di Matteo 1997), according to which good B is produced in the South by a technique that uses E much more intensively than it is used in the North. This translates into a much higher value of D in the South than in the North. This assumption enables us to say that the consequences of the increase in L in the North are greater than those derived from the decrease in L in the South.

The shift in the curve $A^S(N)$ caused by the variation in L is greater than $A^S(S)$, as illustrated in the figure below in which the broken (unbroken) lines indicate supply of good A in the two countries after (before) migration of labour, and P'_B indicates the new equilibrium of the terms of trade.

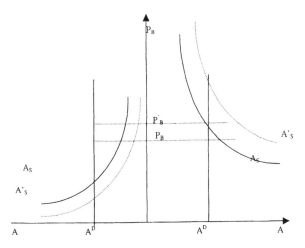

Figure 2: Supply for good A and variations in countries' labour supply

The derivative of supply of A with respect to variations in the labour supply in the two countries, obtained using the theorem of the implicit function is:

$$\frac{\partial A^S(S)}{\partial L^S(S)} = \frac{a_1}{D} < \left(\frac{a_1}{D}\right)N = \frac{\partial A^S(N)}{\partial L^S(N)}$$

The inequality is because D is much greater than D_N under the present assumptions. The terms of trade, or the price of the resource intensive good B, increases and the wage difference between the two areas diminishes.

What happens when the natural resource is extracted? To answer this question we need to know the conditions faced by those working in the resource extraction in the South. In the present context of the South, namely subsistence economy, it is reasonable to assume that the worker has only his labour to exchange for goods. He will therefore work more (and not less) if there is a relative decrease in the price of good B, extracting more of the resource in order to purchase other, let us say "subsistence", goods. In other words, the price of these goods increases in relative terms and the worker works more, not less (as would be the case if q did not increase). However in the situation chosen by the worker, q is equal to the ratio of marginal utility of leisure and marginal utility of "subsistence" goods. Under certain plausible conditions on the utility function

(Chichilnisky 1994, App. B), this ratio increases as the relative price of the goods on which his "subsistence" is based, increases.

Migration of labour from the South increases the price of the good B, leading to extraction of less natural resource. It can also be shown that welfare in the South increases as a result of migration and that a similar effect is obtained by a policy of trade restrictions by the South (Chichilnisky and Di Matteo 1997).

The economics of global warming

The term "greenhouse effect" refers to a natural phenomenon indispensable for life on the planet Earth. Without it, the mean global temperature would be more than 35 C° lower than the present temperature of about 18°C. What worries the international scientific community is not the existence of the greenhouse effect, as commonly thought, but the rapid increase in mean global temperature caused by human activities capable of producing climatic changes on a vast scale.

Many human activities that have developed since the second industrial revolution discharge into the atmosphere substances that have the capacity to alter the atmospheric composition of so-called greenhouse gases, namely gases having the property of maintaining the earth's temperature close to the mean to which human civilisation has been accustomed over the millennia. An increase in the percentage of greenhouse gases in the atmosphere would lead to an increase in volume of these gases, causing an increase in global warming without precedent in the climatic history of the planet, particularly as regards the speed with which it would manifest. This increase in global warming must however be understood, in statistical terms, as an increase in the number of hot days per year in the different climatic belts.

Evaluation of global warming has changed over the years with scientific knowledge of climatic changes in general and the climatic effects of technology applied to production. At the beginning of the 20th century, the increase in greenhouse effect was seen as positive by certain scientists because in their opinion it countered the incipient threat of a glaciation; today the phenomenon is regarded in the opposite way, namely it is feared that the increase in greenhouse effect induced by increasing atmospheric concentrations of substances derived from man-made emissions may lead to new weather patterns (and hence not only an increase in temperature but to overall climatic instability) associated with unpredictable and irreversible changes on a large scale that can no longer be controlled by man.

The main man-made cause of global warming is the increasing quantity of emissions of greenhouse gases, principally due to energy production and industry, but also to deforestation and a complex set of side-effects caused by chemical stress to the atmosphere by various interacting chemical substances.

The evaluation of man-made emissions of greenhouse gases and the release of a particular greenhouse gas (carbon dioxide) associated with deforestation is problematical because of uncertainty about the relations between them and their physical and socioeconomic impact. Greenhouse gases include carbon dioxide (CO_2, which accounts for 55% of the effect), methane (CH4, 15%), certain chlorofluorocarbons (CFC 1, CFC 12, 17%), other CFCs (7%) and nitrous oxide (N_2O, 6%). Carbon dioxide emissions are therefore by far the main source of the increase of this effect. This is particularly evident if we consider the dynamic and cumulative dimension of the emissions. Besides the quantity released to the atmosphere at a given time, it is also necessary to consider the quantity released in the past, because the residence time in the atmosphere may be as long as 100 years.

Man-made emissions of carbon dioxide account for about 5% of total emissions (natural and man-made) and most come from combustion of fossil fuels (natural gas, oil, coal, petrol, wood, coal gas). The impact in terms of CO_2 emissions from fossil fuels is not homogeneous but varies with energy source and combustion conditions.

All greenhouse gases have increased in atmospheric concentration in the last two centuries, and the atmospheric concentration of CO_2 is 346 parts per million (volume). According to certain estimates, this quantity could cause a rise in mean global temperature of 1.5-4.5 C° if 1990 atmospheric concentrations should double. If we also consider that the increased consumption of fossil fuels since pre-industrial times has been accompanied by intense deforestation, we obtain an idea of the entity of the build up of greenhouse gases in the atmosphere.

There is also much uncertainty about the socioeconomic impact, or local effects, of global changes and about man's capacity to respond to these changes. Four aspects can be distinguished:

1) uncertainty about future emissions of greenhouse gases;
2) uncertainty about climatic effects;
3) uncertainty about regional impact;
4) uncertainty about social adaptation.

The different evaluations depend in turn on evaluation of feedback, that is, the role played by reflecting surfaces such as clouds, glaciers and oceans. Feedback may result from an increase in water vapour concentrations caused by the increase in mean global temperature, by a decrease in the earth's albedo due to a decrease in reflecting surfaces (melting of perennial glaciers), by an increase in atmospheric albedo caused by the extra clouds produced by increased evaporation.

To determine whether a "greenhouse problem" exists, it is not only necessary to evaluate physical impact, but also socio-economic effects. We need to know why it is a problem and what the options for solving it are. We need to know what present and future generations will be damaged by climatic change and to what extent. We need to know the entity of irreversible damage and of potential loss of well being. The effects of global warming will mainly be felt by the next generation, whereas the onus of emission control falls on the present generation.

The uncertainty surrounding socio-economic impact has produced some extreme attitudes: certain scholars have stated, with varying degrees of explicitness, that the geographic distribution of the damage to some degree transforms a reciprocal global externality into a unilateral one, in which all countries are responsible for greenhouse gas emissions but the effects are not incurred by all of them. This point of view is based on consideration of only the direct effects, on the conviction that climatic changes induced by an increase in the greenhouse effect are limited in time and space, and on the concept that this will lead to a new equilibrium between countries incurring damage and countries gaining in overall well-being.

However there is no reason to think that climatic change will stop at a certain point. Some sustain that 100 years is too short a period to understand real future effects. Three centuries has been suggested as a more realistic time span, in which it could emerge that the benefits to be gained from reduction of emissions are much more substantial.

With reference to political options, three types of action are possible:

i)   the "do nothing" option (passive or natural adaptation);
ii) the active adaptation option (including the climate engineering option);
iii) the emission abatement option (plus increased emission absorption).

Active or preventive adaptation evaluates scientific uncertainty, minimising future effects of climatic changes by adopting preventive and mitigative measures of a "no regret" character, such as increased defence of seas and rivers, increased investment in research and development, increased expenditure for irrigation, and so forth, with respect to the natural adaptation option.

Unlike options i and ii, that of emission control is directed at the source of the problem. It involves measures leading to the progressive reduction of greenhouse gas emissions, elimination of CFC, replacement of fossil fuels with others that release less greenhouse gases, reduction of methane leaks and carbon monoxide production during combustion, efficiency and energy saving, and progressive use of renewable resources. Unlike the first two options, option iii can only be significant if applied on a global scale. Once this option is chosen in the framework of an international agreement, it is necessary to establish: a) how much abatement; b) the instruments to use to achieve it; c) how to allocate targets for the different countries.

Through knowledge of the overall situation of the various countries, it is nevertheless possible to pinpoint major and minor producers of emissions of greenhouse gases and different capacities of adaptation. These overall situations provide important guidelines in the choice of strategy. Although the industrialisation of developing countries will lead to an increase in emissions of greenhouse gases, these levels are still much lower per capita than in industrialised countries. The United States contributes more than any other country to atmospheric levels of carbon dioxide, with annual emissions amounting to 24% of the total, followed by ex-USSR 17%, the European Union 12.3% and China 11.6%. These four groups of countries alone are responsible for more than 60% of the total emissions.

In the last 20 years, OECD emissions have increased very little, but non-OECD countries, presumably developing countries, emit increasing quantities of CO_2. China, in particular, has emitted increasing quantities of CO_2 in this period, despite low per capita emission levels. This is mainly due to demographic increase: China accounts for 21% of the world population, followed by India with 16%. If we consider that demographic increase in Asia could reach 40% by 2025 and that China and India produce most of their energy from coal, which is the fuel releasing the most CO_2 per unit, it is evident that the development of third world countries in the next 30 years and their energy and demographic policies are fundamental for the definition and negotiation of global abatement strategies. If China and India reached the per capita emission levels of American citizens, world CO_2 emissions would probably triple.

The problem is not to limit emissions and as a result the development of these countries but rather to limit emissions as these countries continue their economic development. Apart from its more obvious advantages, economic development is therefore an objective to defend in the context of global warming. In terms of strategic relationships, for example, the choice of efficiency and energy saving in industrialised countries and LDC operate synchronously: achievement of the former may lead to sustainable development in the latter.

To conclude, the important relations between energy, population and economic development in the context of global warming means that they should be coordinated through comparison of the various national situations and through integration of national and international strategies. This implies tackling global warming in an international context. Whenever these strategies are limited to a national level, there is the risk that special national conditions become special cases which are modified because of other political and economic problems, delaying action and undermining its effectiveness.

The market of marketable permits: efficiency and distribution

Introduction

In this section again we show the importance of the distribution of property rights as far as the problem of decreasing atmospheric concentrations of carbon dioxide is concerned. Atmospheric pollution is traditionally regarded as an example of (negative) externality. To obviate it, a so-called Pigouvian tax on quantities of greenhouse gases emitted or a system of marketable emission permits has been proposed. At this stage of the analysis, the two proposals are equivalent under conditions of certainty, though differences emerge when the question is examined in more detail. In any case, besides theoretical differences, the economic effects of a tax on CO_2 emissions and a system of marketable permits can be demonstrated quantitatively. This is done using computational models of general economic equilibrium (CGE) by which systems consisting of thousands of equations can be solved in a few minutes. We shall not go into detail about how these models work (for details see Shoven & Whalley 1992). They make it possible to evaluate deviations of the main economic variables from their spontaneous trends by implementation of one or more measures of economic policy over a given period of time.

Once it has been decided which of the two to use, on the basis of their merits in solving the problem, it is necessary to decide which countries will jointly apply the tax or set up a global market of permits. This crucial and difficult problem will be discussed in the following sections. Here let us suppose that it has already been solved. For example, the European Union is an example of supranational power that could decide to introduce a tax on CO_2 emissions. The EU did in fact make a proposal of this kind in 1992 after a series of studies on the practical and theoretical aspects (e.g. Carraro & Siniscalco 1993) and the development by the OECD of a computation model "GREEN", to evaluate the effects of the tax (see Lee et al. 1994). The proposal was shelved for reasons that will become clear in the following sections.

In all these discussions, however, the starting point, that is the assumption that CO_2 production is completely similar to an external diseconomy, was never examined critically. It is clear enough that an increase in CO_2 is equivalent to a decrease in the good "clean air". This is a public good because: i) individuals who do not want to pay for it cannot be prevented from consuming it; ii) the fact that an individual consumes it does not decrease the quantity available for others. However, it is an atypical public good. Normally a public good is not produced by private firms by virtue of its non exclusive nature and is typically offered by the public sector. Clean air is a public good which is produced in a decentralised manner by a series of private individuals. This makes it worthwhile revising the conditions of Pareto efficiency of its production from a theoretical point of view.

Clean air: a public good produced privately

Let us first consider the case in which the well-being of *any* country depends on the consumer goods and the clean air that is the indirect result of production of goods by *all* countries. Let us suppose that in each country the value of the difference between consumption and production is zero, i.e. they have to achieve the trade balance. The formal model (a simplified version of that put forward by Chichilinisky et al. 1993) that reflects this hypothesis is as follows.

Each country i ($i=1, 2, ...I$) has the following problem:

Max $U_i(c_i, a)$ subject to:

$$U_k(c_k, a) = N_k \quad k = 1,2,.....I \quad \text{and} \quad k \neq i$$

$$a_i = \Phi_i(c_i), \quad \text{with} \quad \Phi'_i < 0$$

$$\sum_i a_i = a$$

where U_i is the utility function of country i, c_i the consumption of the private good in country i, a_i the quantity of clean air in country i, N_k a given level of utility in country k, Φ_i the production function of clean air in relation to consumption of the private good in country i. The Lagrangian L to maximise for country i is:

$$L = U_i[c_i; \sum_i \Phi_i(c_i)] + \sum_k \lambda_k [U_k(c_k; \sum_i \Phi_i(c_i)) - N_k]$$

The first order condition for country i is:

$$\frac{\partial L}{\partial c_i} = \frac{\partial u_i}{\partial c_i} + [\frac{\partial u_i}{\partial a} + \sum_k \lambda_k \frac{\partial u_k}{\partial a}]\Phi'_i = 0 \quad \text{with } k \neq i$$

or setting $Z = [\frac{\partial u_i}{\partial a} + \sum_k \lambda_k \frac{\partial u_k}{\partial a}]$

$$-\frac{\frac{\partial u_i}{\partial a}}{\Phi'_i} = Z$$

How can we interpret the conditions of Pareto efficiency? The conditions are that the product of marginal utility of consumption of the private good and the marginal cost of abatement of CO_2 emissions must be equal in each country; in other words, the ratio of marginal utility of consumption of the private good to the marginal rate of transformation between the private consumer good and the public good is constant. In fact the marginal cost of emission abatement is simply the reciprocal of the latter quantity: reducing emissions means producing less, in the this case, of the consumer good (for simplicity, assumed to be one). This condition has major implications.

Since the marginal utility of consumption is different from country to country and is presumably greater in poor rather than in rich countries, Pareto efficiency requires that poor countries bear a lower marginal cost of emission abatement than rich countries. If we reasonably assume that the marginal cost of abatement is increasing, then the poor country must reduce a smaller quantity of emissions than it would otherwise have done (as in the figure below).

163

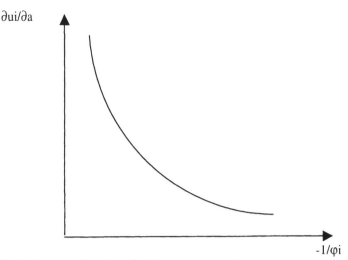

Figure 3: Emissions reductions in the poor country

The second implication is that Pareto efficiency is not achieved by using marketable permits as an abatement measure. A country will find it worthwhile buying permits until the marginal cost of abatement is greater than the price of the permit; however, since there is always only one price of a permit in a competitive market, the marginal cost of reduction is the same in different countries. This contradicts a necessary condition for efficiency in the general case in which the marginal utilities of consumption vary from country to country. Clearly if this were not so (i.e. if the marginal utilities of consumption were the same in all countries), equality of the marginal cost would be a condition ensuring efficiency.

Initial distribution of permits and efficiency

The above suggests that if there were a way to make these marginal utilities equal, the system of permits would be consistent with the requirement of Pareto efficiency. One way is to allow appropriate transfer of the consumer good between countries. When countries are not obliged to balance consumption and production, i.e. when they can consume what they produce plus (minus) what they obtain from (spend for) the sale (purchase) of permits, the system of marketable permits can achieve Pareto efficiency.

In this case, each country has the following problem:

Max $U_i(c_i, a)$ subject to:

$$U_k(c_k, a) = N_k \quad k = 1, 2, \ldots I \quad \text{and} \quad k \neq i$$

$$a_i = \Phi_i(y_i), \text{ with } \Phi'_i < 0$$

$$\sum_i a_i = a$$

$$\sum_i y_i = \sum_i c_i$$

The Lagrangian L to maximise for country i is:

$$L = U_i[c_i; \sum_i \Phi_i(y_i)] + \sum_k \lambda_k[U_k(c_k; \sum_i \Phi_i(y_i)) - N_k] + \theta[\sum_i y_i - \sum_i c_i]$$

The first order conditions for country i are:

$$\frac{\partial u_i}{\partial c_i} = \lambda_k \frac{\partial u_k}{\partial c_k} \text{ with } k = 1, 2, \ldots I \text{ and } k \neq i$$

$$\Phi'_i [\frac{\partial u_i}{\partial u_a} + \sum_k \lambda_k \frac{\partial u_k}{\partial a}] = -\theta$$

Consistency requires that $\lambda_k = 1$. The equality of marginal costs is therefore a necessary condition for efficiency: note that the term in square brackets is a constant, Z, as in the previous section.

The necessary redistribution can be done by appropriate initial distribution of pollution rights; in fact a quantity of permits in excess of the current pollution level can be sold on the market, obtaining funds to buy the private good. This scheme of assigning permits is different from others, such as the often proposed scheme of a quantity of permits proportional to the population.

This is a further example that the distribution of property rights on the good, clean air, is important for achieving efficiency. In this case the separation between efficiency and the equity so dear to the tradition of the economics of welfare, breaks down.

Some difficulties in instituting a market for tradeable permits

The approach in terms of distribution of property rights can at most be used heuristically but not in terms of economic policy, without some idea about how to solve the problems arising when this new market is instituted. It involves decisions by the countries that have to reach an agreement on the global quantity of permits to be given and on their distribution. Institution of the market has high costs, and many technical details on the working of the market have to be established. The first question (decisions of countries) will be dealt with in a later section. With regard to the second, we can think of market formation, transactions and monitoring costs. The former are essentially fixed costs and are derived from the construction of a normative framework defining the types of contracts and goods to be exchanged, and the methods of payment.

The form of the marketable permit is important: not only must it specify the amount of CO_2 that can be emitted by the holder, but also the duration of the permit. For example, the holder may be authorised to emit a quantity X of CO_2 in 10 years. The distribution of the quantity in the 10-year period is left to the discretion of the holder because what counts as far as the greenhouse effect is concerned is the overall amount of CO_2 in the atmosphere at a given time. However the length of the period is just as important for policy makers as it is for traders.

For decision makers it may limit the possibility of varying the number of permits with changes in the total quantity of atmospheric CO_2 admissible. This limit also depends on scientific understanding of the problem. Likewise for permit holders, the implementation of production plans is conditioned by the length of the permit: once the permit has been purchased, the cost of emitting CO_2 does not change for the period of its duration.

The cost of transactions on the market depends in an inverse manner on the size of the market; in general, the larger the market, the easier it is to find the counterpart. The greater the number of traders, the greater the competition; this is facilitated by the participation not only of states, but also of firms and other economic agents. The market functions even better if a forward market is introduced, or even, as suggested by Chichilnisky (1996), by the institution of an international bank which lends and borrows permits.

Finally, there are surveillance and enforcement costs. The former depend on how dispersed are the sources of production of oil, coal, etc. where most CO_2 emissions are concentrated. These costs can be limited by checking that producers do not emit more than is authorised. For holders who do not observe the rules, there is the problem of sanctions, which (at least in principle) is easily solved within single states, but becomes much more complex when it come to punishing a state in the absence of a world organisation with supranational powers. In the absence of the latter, observation of the limits defined by the permits could be facilitated by the fact that it is also in the interests of the country itself.

As we shall see, this problem is closely related to the problem of participation of countries in the permit market, or more generally in agreements on emission abatement. We now examine this aspect.

International cooperation versus free-riding

From the point of view of environmental and socioeconomic damage, the size of the externality and the underlying cause determine the number of states involved and the distribution of damage and benefits derived from the externality. If the externality caused by environmental damage is mutual, the states are incentivated to act, especially if prompted by internal forces. Unilateral action, in the case of mutual global pollution, can be seen as an implicitly cooperative solution, having global benefits provided the other polluters do not counterbalance the reduction of emissions of the country with an increase in their own. The following results are from Barrett (1994).

Let us consider the following hypotheses:

1) N identical countries, $i=1, ..., N$;
2) each country emits the same quantity of pollution ex ante;
3) each country has the same pollution abatement costs and the same benefit functions;
4) the benefit and marginal cost curves are linear.

Each country also has a benefit function derived from *global* abatement of emissions, Q, with $Q = \Sigma_i \, q_i$, , where q_i is the level of abatement of emissions in the *ith* country, namely:

$B_i = b(aQ - Q^2/2)N$

and an associated cost function of abatement of *its own* emissions:

$C_i(q_i) = cq_i^2/2$

The total net benefits of the *ith* country are:

$\pi_i = B_i(Q) - C_i(q_i)$

and net benefits of all countries are:

$\Pi = \sum_i \pi_i = \sum_i B_i(Q) - C_i(q_i)$

If there is full cooperation between the N countries, the first best solution is given by the value of Q for which Π is maximum:

$Q_c = \text{argmax } \Pi$

In other words, the full cooperation solution requires each country to balance its marginal cost curve (*MCi* in the figure below) with its global marginal benefit curve (GMB).   Global and individual levels of cooperative equilibrium are then:

$$Q_c = \frac{aN}{N + \dfrac{c}{b}}; \qquad\qquad q_c = \frac{a}{N + \dfrac{c}{b}}$$

where Q_{nc} is the optimum level of global abatement and q_{nc} the quantity of individual abatement associated with cooperative equilibrium.

In the absence of cooperation, each country chooses its own quantity, in a context à la Cournot, assuming as given the choices of the other countries.   In other words, each country chooses the level of abatement that maximises its net individual benefits of abatement, so that its costs and marginal national benefits are equal (the curves MCi and IMB in the figure).

The global and individual levels of non cooperative equilibrium are:

$$Q_{nc} = \frac{a}{1 + \dfrac{c}{b}}; \qquad\qquad q_{nc} = \frac{a}{N(1 + \dfrac{c}{b})}$$

where Q_{nc} is the optimum quantity of global pollution abatement in the non cooperative case and q_{nc} qnc the individual quantity of abatement associated with non cooperative equilibrium.

Since $Q_c > Q_{nc}$, each country would be better off from the environmental point of view with the full cooperation solution, but they are not incentivated to achieve the level of pollution abatement associated with cooperative equilibrium, unless bound in some way to do so.

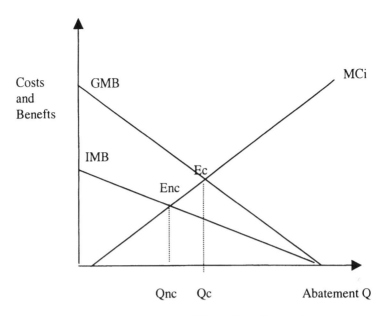

Figure 4: Private vs. collective costs and benefits of emissions reduction

See:Barrett (1990a)

If all countries have different national benefit costs, the sum of the abatements undertaken individually could lead to less than optimum global abatement. However, we assumed that the N states are identical, so that the net global benefits will be the sum of net individual benefits. Maximisation of this sum implies that each country chooses the level of abatement for which individual marginal costs equal global marginal costs.

For a given N, the difference between Q_c and Q_{nc}, i.e. between the full cooperation and non cooperation solutions, will depend on the slope of the cost c and benefit b curves. The difference between the two equilibrium levels will tend to reflect the size of the ratio c/b.

The following analysis is made in terms of a "one-shot" game in which the players meet only once. This hypothesis is said to be unrealistic and it has been pointed out that the possibility of future meetings induces players to consider how strategies evolve in time. Actually, however, the countries meet repeatedly before coming to an agreement and after some kind of agreement has been made.

Repetition of the game could in fact transform it in terms of c/b and the number of countries involved. An important element for the choice of cooperative or non cooperative strategies is in fact the number of countries involved. As the number of countries increases, the total sum of marginal

169

benefits derived from abatement policy increases, and with it, the divergence between the cooperative and non cooperative solutions. However, as the number of countries increases, the importance of the marginal contribution of individual countries decreases. This may induce single players not to limit pollution, or to limit it less than they would otherwise have done.

Hence there seems to be a trade-off between the number of parties to an international agreement and each country's incentive to cooperate fully.

This assumption is worth examining because it seems to contradict the widespread idea that if an agreement is not global, it is unlikely to be effective. The experience of the Conferences of Rio and Kyoto, however, suggest otherwise.

A last question is what happens in this model if we no longer assume that all countries are equal. The analysis is then complicated by the fact that in order to make the sum of national marginal benefits coincide with that of global marginal benefit, the countries that benefit most from abatement undertaken by the others must agree to make appropriate side-payments to reincentivate single countries to limit their emissions. If this assumption is not made, it seems difficult to achieve optimum global equilibria unless the agents cooperate.

In the next section we look at the sustainability of agreements when countries are not identical, and hence the problem of side-payments, and the sustainability of coalitions in a repeated context, where the coalition may undergo defections or acquire new participants.

Agreements between states: the optimum number of parties to a coalition

In the previous example, let us consider S identical states, in the group of N states, that agree to cooperate on pollution abatement, whereas the other P states do not cooperate. A leadership phenomenon à la Stackelberg may occur between S and P, with S countries acting as abatement leaders and P countries as abatement followers. The former decide on targets in the framework of their agreement and the latter follow this strategy in a non cooperative way.

If the strategies of S and P are coherent in time, according to Barrett (1989) this will lead to a stable agreement and a better overall result than if all countries signed the agreement. In fact if P did this it would initially reinforce the agreement, but the corresponding increase in per capita reductions would make it necessary to rewrite the agreement. Countries S

will sign it only if the increases in benefits of each state, determined by the new members, exceed the increase in abatement costs. If this does not occur, there are probably defections from the agreement which ends up being weakened by extension of the cooperation. In other words, *the optimum number of parties to a cooperation is not absolute but depends on the number of countries participating and the number not participating, i.e. on the spatial and temporal distribution of the damage and its perception.*

Returning to the previous model, if N is large, new memberships and defections will change little and the agreement will not give satisfactory results. Besides, as N increases, the incentive to free-ride increases (because of the marginal position in which each country finds itself), as long as the advantages of cooperating are greater than any sanction against free-riding.

For a given N, the number of parties will increase if c/b decreases; but if c/b is small, it means that the cost of abatement is also small relative to the benefits, in which case the benefit of an international agreement takes on a new aspect.

It often happens that many states sign international agreements when the cooperative and non cooperative solutions are not very different. This is because a state is more likely to participate in the negotiations when it already has some initiative in mind. It may also happen that nearly all countries, except those most responsible for the international good to be safeguarded, sign an international agreement. An example is the *International Whaling Commission* of which all countries of the world were part except the three that carried on whaling.

Clearly the ratio c/b, the number of countries and the type of incentivation are what distinguishes a cooperative or non cooperative solution.

Let us take the previous model and suppose that a fraction a of countries sign an international agreement to reduce emissions so as to maximise net benefits for the group. These countries act as leaders in the sense that they decide collective abatement levels and wait to see the reaction of the other countries. Countries that do not sign the agreement, $(1-a)$, behave *à la* Cournot, in that they decide their own abatement levels, assuming as given emission abatement levels of the other countries. If q_a is the emission abatement of signatory country a, and qN that of non signatory country N, the solution can be written:

$$q_a(a) = aNh(a)/[c/b + a^2Nh(a)]$$

$$qN(a) = [\ (c/b)/(c/b + 1 - a)]/[c/b + a^2 Nh(a)]$$
where $h(a) = [(c/b)/(c/b + 1 - a)]^2$.

We still have to determine a^*, namely the equilibrium number of the signatory countries. Let us suppose that P_a is the net benefit of signatory country a and P_N that of non signatory country N. If:

$$P_N(a - 1/N) \leq P_a(a)$$

then the agreement is "internally stable" and the participating countries do not defect. Although defection implies a reduction in abatement levels and costs for the defecting country, it weakens the agreement and the consequent loss of benefits exceeds the reduction in costs (of the defecting country), producing a net loss for that country.

If on the other hand:

$$P_N(a) > P_a(a + 1/N)$$

the agreement is "externally stable" and no external country will want to participate. Although entry of an external country would increase global benefits, it also increases costs for that country, inducing it to stay out. An agreement is stable if it is internally and externally stable at the same time.

All other things being equal, this problem is solved by determining the optimum number of parties a^*. Barrett does this using an iterative algorithm from a value $a^* = 0$. The condition of stability is not, however, an objective to achieve at any cost.

The model of Carraro and Siniscalco develops Barrett's (1990, 1990a) analysis of non cooperative strategy, comparing it with various cooperative strategies. The crucial point is evaluation of the elasticity of the reaction curves of the countries.

Carraro and Siniscalco state, however, that the condition of stability does not itself solve the problem of optimum management of transnational commons. They demonstrate that in the previous model, stable coalitions can generally be formed by $j^* < N$ countries, where j^* is a small number (between 2 and 4) independent of the value of N. The problem is to see how far the coalition can be expanded without losing stability or benefits but gaining in terms of environmental efficacy. The only way to expand a stable coalition is to change the incentives for external countries by appropriate transfer of self-financed utility of the coalition. It is not, however, possible to expand a stable coalition by transfer of self-financed utility without weakening the agreement.

According to Carraro and Siniscalco, the only way to avoid this is to introduce a minimum degree of commitment between members of the initial coalition, or some of them.

Transfer of utility can also be financed by countries external to the coalition, which prefer to finance the reduction of emissions in other countries because of a different structure of abatement costs. This possibility is called "external support", and in the model of Carraro and Siniscalco it is demonstrated that under certain conditions, the number of countries forming a stable coalition can be doubled or tripled with increasing environmental efficacy. Self-financing with commitment and external support are therefore two ways of effectively and efficiently expanding stable coalitions between few countries.

REFERENCES

Barrett, S., 1990a, The Problem of Global Environmental Protection, Oxford Review of Economic Policy, VI,1, pp. 68-79.

Barrett, S., 1990b, International Environmental Agreements as Games, mimeo Beltratti, A., ed., 1996, Models of Economic Growth with Environmental Assets (Dordrecht: Kluvier Academic Publishers).

Bromley, D.W., 1992, The Commons, Common Property, and Environmental Policy, Environmental and Resource Economics, II, pp. 1-18.

Carraro, C., ed., 1994, Trade, Innovation, and Environment (Dordrecht: Kluvier Academic Publishers).

Carraro, C., and Siniscalco, D., 1991, Environmental Innovation Policy and International Competition, CRPR, discussion paper 525.

Carraro, C., and Siniscalco, D., 1992, The International Dimension of Environmental Policy, in European Economic Review, pp. 379-387.

Carraro, C., and Siniscalco, D., 1993, Strategies for the International Protection of the Environment, Journal of Public Economics, LII, pp. 309-28.

Carraro, C., and Siniscalco, D., 1993 The European Carbon Tax: an Economic Assessment (Dordrecht: Kluvier Academic Publishers).

Chichilnisky, G., 1994, North-South Trade and Global Environment, American Economic Review, LXXXIV, pp. 851-874.

Chichilnisky, G., 1996, Development and Global Finance: The Case for an International Bank for Environmental Settlements, New York, UNDP, Office of Development Studies, WP10. pp. 48,

Chichilnisky, G. and Di Matteo, 1997, Trade, Migration, and Environment: a General Equilibrium Analysis. In Chichilnisky,G., Heal, G.M., and Vercelli, A. eds., Sustainability: Dynamics and Uncertainty, (Dordrecht: Kluvier Academic Publishers).

Chichilnisky, G., Heal, G.M., and Starrett, D.A., 1993, International Emission Permits: Equity and Efficiency, Stanford University, Centre for Economic Policy Research, Technical Paper n.381.

Gandolfo, G., 1994, Corso di Economia Internazionale, Vol. I, (Torino: Utet Libreria) Global Climate Change: a Symposium, 1993, Journal of Economic Perspectives, VII, 4, p. 3-86.

Grubb, M., 1989, The Greenhouse Effect: Negotiating Targets (London: The Royal Institute of International Affairs).

Hardin, G., 1968. The Tragedy of the Commons, Science, 13/12/1968.

Hoel, M., 1990, Efficient International Agreements for Reducing Emissions of CO_2 Memorandum from Department of Economics, University of Oslo, n. 6.

Hoel, M., 1991a, Emission Taxes in a Dynamic International Game of CO_2 Emissions, mimeo.

Hoel, M., 1991b, Global Environmental Problems: The Effects of Unilateral Actions Taken by One Country, Journal of Environmental Economics and Management, XX, pp. 55-70.

Hhoel, M., 1992, Carbon Taxes, An International or Harmonized Domestic Taxes?, European Economic Review, XXXVI, pp. 400-406.

International Panel on Climate Change (1995), Summary for Policy Makers of the Contribution of Working Group I-III to the IPCC second report, (Cambridge: Cambridge University Press).

Jaffe, A. B., Peterson, S.R., Portney, and Stavins, R., 1995 Environmental Regulation and the Competitiveness of US Manufacturing, Journal of Economic Literature, XXXIII, pp. 132-163.

Onufrio,G. and Gaudioso, D., 1989, L'atmosfera avvelenata, (Padova: Franco Muzzio Editore).

Pearce. D., Baber, W. And Markandya, A., 1990, Sustainable Development: Economics and the Environment in the World, (Aldershot: Elgar Publishing Company).

Porter, M.E., and van der Linde,C., 1995, Towards a New Conception of the Environmental-Competitiveness Relationship, Journal of Economic Perspectives, IX, 4, pp. 97-118.

Rauscher, M., 1997, *International Trade, Factor Movements, and the Environment* (Oxford: Oxford University Press).

Shoven, J.B. & Whalley, J., 1992, *Applying general Equilibrium* (Cambridge: Cambridge University Press).

SO2 Trading: a Symposium, 1998, *Journal of Economic Perspectives*, XII, 3 pp. 53-88.

Weyant, J.P., 1993, Costs of Reducing Global Carbon Emissions, *Journal of Economic Perspectives*, VII, 4, pp. 27-46.

World Bank, 1992, *Word Bank Report: Development and the Environment* (Oxford: Oxford University Press).

World Commission on Environment and Development, 1987, *Our Common Future* (Brundtland Report) (Oxford: Oxford University Press).

PART THREE
EVALUATION PROCESSES AND POLICY CHOICE

7. Exploring Biophysical Approaches to Develop Environmental Taxation Tools:
Envitax, to Face the "New Scarcity"

Salvatore Bimonte and Sergio Ulgiati

Introduction

Recent debate on environmental taxation and proposals for Carbon/Energy taxes; the Delors White Paper on "Growth, competitiveness and employment" (European Commission, 1993) based on the double dividend idea; OECD recommendations inviting member countries to make a greater use of economic instruments (OECD, 1991) and to mutually reinforce environmental and fiscal policies (OECD, 1993), has attracted renewed attention to economic instruments as suitable and efficient tools to achieve environmental goals.

It has been shown that economic instruments can achieve these goals at a lower cost than command and control policies (static efficiency). They can also provide a dynamic incentive for emissions abatement and technological innovation (dynamic efficiency). Here we are not concerned with the "efficiency" argument for taxing pollution, for which the reader is referred to the large body of literature on the theoretical aspects of environmental taxation (for example, Baumol and Oates, 1988; Bimonte, 1995). The aim of this paper is to discuss the following issues. Given the multidimensionality and the holistic nature of the environmental problem, it is maintained that human society is now facing a "new scarcity", as defined below. It is suggested that a new and more comprehensive taxation scheme than those proposed at different levels is needed. In view of the ongoing debate on correct utilisation of environmental tax revenues, it is suggested that ecological and economic sustainability can be better addressed by a taxation scheme that emulates natural processes, namely by feeding back natural capital. To this end, the paper proposes:

1) to tax activities according to their quality attributes, where "quality" means demand for environmental support. This can be achieved by linking the tax tool with new and more comprehensive physico-chemical indicators of environmental dynamics, based on the emergy accounting methodology (Odum, 1996);

2) to consider the environment as a life supporting system as well as a fund rather than as a stock.[1] Environmental performance should be preserved through feed-back processes aimed at restoring the productive infrastructural functions of the biosphere.

In Section 2 we define the problem and give the definition of "new scarcity"; in Section 3 we introduce emergy accounting methodology and the maximum power principle, defining the indicators according to which activities should be taxed; in Sections 4 and 5 we formalise the structure of the new taxation scheme; Section 6 describes some case studies that illustrate how different indicators affect production and consumption paths.

Defining the "new scarcity"

The decreasing life supporting capacity of our planet will be the major concern of the third millennium. We refer to it as the "new scarcity". It encompasses scarcity of the environment as a whole, considered as a resource made up of many different resources (fertile topsoil, clean air and water, diverse ecosystems, etc.). In hierarchy theory as applied to complex systems it is well known that the properties of the whole are often different from the properties of the component parts. For instance, if excessive exploitation of fertile topsoil by modern agriculture increases soil erosion, it is not just the ability of soil to support agricultural production that is lost, but also the turnover processes of

---

[1]   A flow is an element that either is only "consumed" or only "produced" by the process; it is either only an input or only an output. It stems from a stock. A fund is both an input and an output; more precisely it is a factor whose economic efficiency is maintained by the very process in which it participates (Ricardian land provides the clearest example of the concept of fund). (Georgescu-Roegen, 1986)

chemicals, micro-organisms and organisms, and the water cycle (evapotranspiration, runoff, deep ground water). These in turn affect many other ecosphere processes in a web of direct effects and feedback. Thus, a much greater loss of stability and homeostatic properties at a higher level are the consequence of weakening a component at a lower hierarchical level.

The "new scarcity" concept is therefore: "general" as it involves all kinds of environmental resources, "organic" as it affects the vital processes of the biosphere, "global" as it concerns the planet as a whole, "irreversible", because environmental processes are governed by thermodynamic law, "uncertain", because our knowledge is intrinsically imperfect.

The greenhouse effect is another example which encompasses many of the above features. It is an irreversible phenomenon since the residence time of greenhouse gases (CO_2, N_2O and CFCs) in the atmosphere may be longer than the time scale of human life; it is global, affecting the planet as a whole; it is characterised by uncertainty, since, despite a significant warming trend, and a general consensus that atmospheric CO_2 has increased due to the combustion of fossil fuels, the link is still far from proven (IPCC, 1995).[2] However, the greenhouse effect is not the only concern. Energy and/or carbon emissions are only part of the problem. What really matters is the life supporting ability of the environment as a whole. This issue can be addressed if we consider processes, rather than their outcome, globally and quality rather than quantities.

Emergy accounting and the maximum power principle

The study of ecosystems and self-organising systems suggests principles by which energy flows generate hierarchies. Energy laws controlling self-organisation are principles quite beyond classical energetics, involving generalisation of the concepts of evolution in energy terms. During the trials and errors of self-organisation, species and relationships are selectively reinforced by more energy becoming available to designs that feed products back into increased production. This phenomenon is the well known maximum power principle (Lotka, 1922a,b; Odum and

---

[2] Nordhaus (1991) suggests that the best investment today would be in learning about climate change rather than in preventing it. In our opinion, the former does not preclude the latter.

Pinkerton, 1955; H.T. Odum, 1983; H.T. Odum, 1996) which suggests that systems that develop and prevail are those that increase and take maximum advantage of available resources. Many authors have investigated the role of energy and thermodynamic conversions in self-organisation processes supporting life,[3] but the quality of the different inputs and output flows has often been ignored, and the same value thus assigned to joules of input originating from processes at very different hierarchical levels (for instance, 1 joule of photosynthesised matter cannot be equated to 1 joule of direct solar radiation, as photosynthesis concentrates solar energy with 0.1% efficiency, therefore requiring a greater quantity of solar radiation).

To account for quality in modelling complexity, an emergy theory was developed by H.T. Odum in the eighties (Odum, 1988; 1996). In short, *solar emergy* is defined as the sum of all inputs of solar energy (or an equivalent amount of geothermal or gravitational energy) directly or indirectly required by a process to provide a given product. Usually these inputs are yielded by another process (or chain of processes), by which solar (equivalent) energy is concentrated and upgraded.

On a unit basis, one joule or gram of a given item is produced by dissipating a certain amount of solar equivalent energy. The amount of input emergy per unit output energy is called *solar transformity*. The solar transformity gives a measure of the convergence of solar emergy through a hierarchy of processes or levels; it can therefore be considered a "cost" factor, a measure of the global process supporting that particular flow. Thus, the total emergy of an item can be expressed as:

solar emergy = amount of an item × its convergence factor (transformity).

The item can be expressed in many different units (joule, gram, mol, kcal, $). The solar emergy is usually measured in solar emergy joules (sej), and the unit for solar transformity is solar emergy joules per joule of product (sej/J). Sometimes emergy per mass of product is also used (sej/g).[4] According to the efficiency of the processes along a given

---

[3] The reader is referred to a recent review of these authors in Jorgensen (1997).

[4] *Emergy is not energy.* The emergy of a given flow or product is, by definition, the total amount of solar energy that is directly or indirectly required to generate the flow or production process. This amount of solar energy is provided by the work of self-organization of the planet as a whole. Part of this work was performed in the past (think of fossil fuels) over millions of years. Part is the present work of self-organization of the geo-biosphere.

pathway, more or less emergy may have been required to reach the same result.

Transformities are calculated by preliminary studies on the production process of a given good or flow. Many of them can be found in the scientific literature on emergy (Odum, 1996). When a large set of transformities is available, other natural and economic processes can be evaluated by calculating input flows, throughput flows, storages within the system, and final products, in emergy units. After the emergy flows and storages of a process or system have been evaluated, a set of indices and ratios useful for policy making can be calculated.

Emergy based indicators of efficiency and environmental loading

Emergy analysis has provided some indices closely linked to environmental issues. Briefly, they are:

Transformity as an indicator of "quality". By definition, transformity is a measure of the emergy convergence through a process, a measure of the intensity of work performed by the environment to "manufacture" a product. Transformity indicates the quantity of the environmental contribution (direct and indirect) to the process. The fraction of past and present environment supporting a process should be taken into account when planning, promoting and taxing economic processes. A low transformity indicates a low "space-time" convergence of the environmental work. If a large amount of work goes into a resource, a large amount of work is also required to replace it. High transformity goods are more difficult to replace and should not be wasted.

The higher the flow required, the higher the present and past environment exploited to support a given process. Thus emergy can be regarded as "energy memory", i.e. the memory of the energy that has gone and is going into the process.

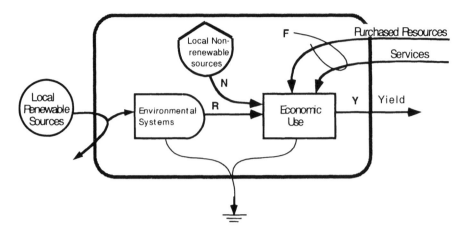

Yield (Y) = R+N+F
$\Phi_R=$ Fraction Renewable = R/(R+N+F)
Emergy Yield Ratio = EYR = Y/F
Environmental Loading Ratio = ELR = (F+N)/R
Emergy Sustainability Index = ESI = EYR/ELR

Figure 1: Aggregated diagram of renewable, nonrenewable and purchased (invested from outside) emergy inputs to an economic system.
Illustrating computation of some emergy based indices and use ratios (Brown and Ulgiati, 1997).

Emergy Yield Ratio (EYR) is the ratio of the total emergy content of the yield to the emergy content of the economic resources invested in the process (Figure 1). A high emergy yield ratio means a high return on the investment, due to the exploitation of locally renewable and non-renewable resources. An EYR equal to 1 means that the process contributes no more than it receives (in terms of emergy) to the economy. Raw and primary resources (fossil fuels, minerals, fresh water, soil, wood, fish, crops, etc.) usually have a high EYR. Manufactured goods have a low EYR, often close to one. This is because the fraction of natural "unpaid work" in raw materials is large, whereas its weight decreases in high manufactured goods.

Ulgiati et al. (1995) suggested that emergy inputs invested by the economy to recycle pollutants and to repair damages to natural and social assets, due to the use of a given resource, should also be accounted for in the EYR formula. They introduced the concept of "emergy yield ratio of end use", which is sometimes much lower than the "emergy yield ratio of a source".

Environmental Loading Ratio (ELR)   is the ratio of emergy inputs under human control (from the economic process; e.g. fossil fuels, labor, etc) to the emergy of natural inputs (sun, rain, etc) (Figure 1). Equivalent definitions are: the ratio of slowly renewable to rapidly renewable emergy sources, or the ratio of stock-limited to flow-limited emergy sources. A high ELR indicates environmental stress and low ecological sustainability.

Emergy Sustainability Index (E.S.I.).   Large EYRs indicate a large contribution to the economy (high economic sustainability in terms of emergy). If this is obtained with a large ELR, it means low ecological sustainability. A low ELR with a low EYR means ecological sustainability but low economic sustainability. Environmental integrity is promoted when yield is maximum and environmental loading is minimum.

Economic and ecological sustainability are therefore linked, and "unified sustainability" can be seen as the ability to contribute to society with the least possible misuse of resources. This can be expressed by a new emergy based index, the ESI, introduced by Brown and Ulgiati (1997):

$$ESI = EYR/ELR = \eta + [\eta^2/(1+\theta)] \qquad (1)$$

where η and θ are the ratios of locally renewable, R, and locally non-renewable, N, emergy inputs to purchased investments, F, provided by the economic system respectively. Application of this index to selected production systems (see Tables 1 and 2, as well as Brown and Ulgiati, 1997) seems to confirm the idea of an aggregate measure of environmental and economic sustainability.

Relevance of emergy accounting to economics

Maximizing emergy through-flow consistent with existing bio-physical constraints is the goal of self-designing systems (Maximum Power Principle). Environmental policies usually focus on a selected aspect of the economic process and optimise with respect to it.

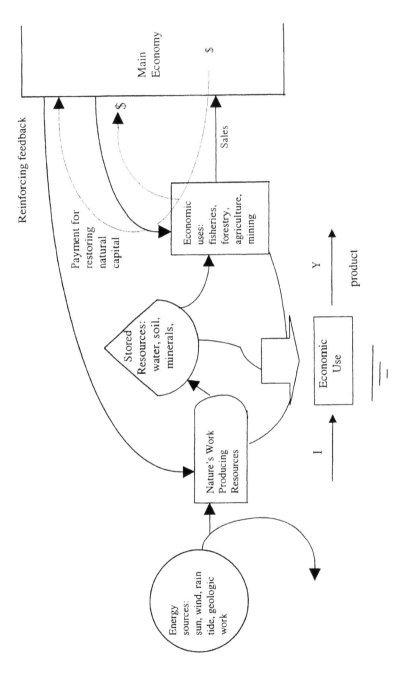

Figure 2: **Model of resources building and use** Resources provided and stored by nature's work are then exploited for economic uses. A reinforcing feedback from consumers is needed to avoid the depletion of natural capital (Odum, 1994).

The same is true of the environmental taxation policies (e.g., energy, carbon and emission taxes) proposed so far. Because they focus on a particular goal, taxation policies do not take a general and global view of the environment's contribution and the basic issue of ecosystem integrity is not addressed. For example, the aim of the carbon tax is to reduce carbon dioxide emissions and avoid global warming, but temperature is not the only resource that requires control or feedback from the economic system. The integrity of the environment as a whole is of fundamental importance for the survival of the human race. Economic instruments, like taxation, must evolve in line with these objectives.

Recent developments in energy analysis, applied to self-designing systems by Howard T. Odum's eco-energetic group at the University of Florida, seem to offer the right thermodynamic basis for new types of taxation involving the restoration of natural capital (Figure 2). Policies that reduce the emergy available to a system or its feedback, generally reduce system vitality, e.g. policies that restrict the use of primary energy at the beginning of the production chain, without distinguishing between different uses of available resources:

> Whereas energy conservation in the sense of increasing efficiency of use has net benefits, an economy that conserves, in the sense of restricting fuel use, tends to reduce its emergy availability and thus its ability to compete economically. Taxing fuels is sometimes offered as an incentive for energy conservation, but reducing fuel has a negative amplifier effect on the economy that may be greater than the increases in efficiency. If the tax reduces luxury and waste, the effect is beneficial (Odum, 1996).

To be effective towards environmental integrity, a taxation policy should therefore focus on the quality of a production process, i.e. on its ability to reinforce the productive basis and avoid waste of resources. Energy is not the only input, nor even always the most important. Emergy theory makes it possible to account for inputs other than energy on a common basis.

The taxation approach to the environmental problem

As stated above, taxation policies have usually focussed on a specific aspect, ignoring the general and quality side of the environmental issue. Eco-taxes have been proposed for products, the manufacture, consumption and disposal of which cause pollution. Examples are taxes on packaging, fertilisers, pesticides, and so on. In most cases, these taxes are intended to tackle specific environmental issues on a case-by-case basis (OECD, 1997a).

Moreover, the introduction or alteration of environmental taxes is often carried out in a context of revenue neutrality. In order to guarantee revenue neutrality, it has been proposed to use tax revenues to cut down other distortionary taxes, for instance VAT (Value Added Tax), so as to leave the total fiscal burden unchanged. This would also enable inflationary consequences to be neutralised (Baker, 1992). This approach is somehow based on the assumption that the environmental tax is distortionary (OECD, 1994). It has also been proposed to reduce labour taxes, such as employers' social security contributions, financing them with a carbon/energy tax, in order to make labour comparatively more economic than fossil fuels. This so-called double-dividend, according to its proponents, would guarantee a reduction in carbon emissions and an increase in employment (Oates, 1991; Pearce, 1991, among others).

This approach burdens taxation with tasks it is not qualified for. The explicit aim of an environmental tax is to decrease emissions and pollution.

The decrease is normally larger in the long run than in the short run. In the long run there is more scope for developing new products and technologies, and changing behaviour patterns. This leads to a reduction in the tax base (i.e. less pollution), which in turn reduces the tax revenues.

We strongly believe that tax revenue utilisation is a critical matter in every assessment related to environmental policy (Piacentino and Smith, 1994). Tax revenue use may affect welfare and the environment. With regard to the former, once a tax has been levied, tax revenues can be used to make some people better-off without making anybody worse-off (Pareto improvement). With regard to the latter, wrong use of tax revenues can exert detrimental feedback on results of environmental policies. This should be carefully investigated. For example, in the case of double-dividend theory, the link between employment increase and final demand should be evaluated. It is not clear from theory whether or not a tax shift from labour-based to an energy-based production process promotes employment; this may depend on many related economic factors (e.g. flexibility of labour market; bargaining strength of employees and employers; the extent to which those outside the labour force bear the tax burden).[5] However, even under the favourable hypothesis, it has been demonstrated that these policies are beneficial for unskilled unemployed. If we assume, as usual, that low income households have a higher marginal

---

[5] On this argument several simulations have been carried out, producing different results. See, among others, Majocchi (1996) and OECD, (1997b).

propensity to consume, we can expect an increase in the final demand. This could partially offset the expected reduction in polluting emissions. Of course, the results depend on many interrelated factors: the marginal propensity to consume of the low income household compared to that of high income households; the extent to which non-wage incomes bear the burden of the tax; the share of energy expenditure on total income; the level of unemployment benefits and so on.

The proposal of an emergy earmarked tax: the Envitax

The need for a more comprehensive and coherent environmental policy, and the problems of traditional approaches to taxation lead us to focus on the quality aspects of processes and propose a new kind of taxation which we call Envitax. It is an emergy earmarked tax. It is based on the belief that environmental taxation is not distortionary[6] and that is more effective if managed in an integrated way. To make economic processes viable over time, sound ecological policy should use tax revenues in positive feed-back actions to the environment.[7]

Products should be taxed according to their demand for environmental services and their global sustainability. The revenues of this taxation should then be used to strengthen the environmental basis of the process, so as to avoid depletion of environmental storages and maximise emergy flow through the system. Emergy based indices provide an useful basis for a taxation policy to discourage misuse of high quality resources and to reward systems that apply a reinforcement feedback to their productive basis (Figure 2). Envitax would enable the policy maker to tackle the "new scarcity".

In our opinion, the "new scarcity" goes far beyond the problem of externalities. In the "new scarcity" context, like under the carbon tax prescriptions, the goal of the tax is no longer to reduce the activity level to the economically efficient point. The aim is now to single out a (relatively) efficient path to follow in the dynamic evolution of processes. Tax should be the economic "carrier" of the scientific information.[8] It enables a degree of

---

[6] Much theoretical literature exists on this topic. For example, see Baumol and Oates (1988).

[7] An example is contribution to special environmental funds, such as waste water charges in several countries (e.g. France).

[8] From this point of view, tax is beneficial *per se*, because it induces greater efficiency in resource utilisation. Technological systems tend to react asymmetrically to price changes.

flexibility in sequential decision procedures. "Flexible positions are not so important because they are safe stores of value, but because they are good stores of options" (Jones and Ostroy, 1984, p.14). Its aim is not to stop unpredictable processes, but to make them manageable in accordance with available information. By distributing decision making over time and adopting a prudent approach at the beginning, it may be possible to gather new, albeit not complete, information about prominent variables. This is equivalent to a reduction in uncertainty of future decisions. Tax can be regarded as the cost to be sustained for reducing uncertainty.

Structure and effects of an Envitax

When the maximum power principle is applied (i.e. a strategy of maximisation and reinforcement of throughput flows by maintenance of available fund) to an economic process it emerges that tax should only be levied at the final steps (the product) and not at the beginning (input flows). What really matters is appropriate use of production factors and services in the process. The "surviving" process would be the one that optimises its efficiency for maximum power output, rather than the one that consumes less of a given factor. If we levy a tax on inputs to prevent a process from using those specific factors we may obtain an unexpected result from the sustainability point of view. The process may have used this factor at optimum "maximum power efficiency", yielding more output. For urgent problems, such as the depletion of ozone layer, the use of specific factors (such as chlorofluorocarbon compounds) may be disincentivated. However, in a global and long term scenario this bit-by-bit strategy may not be the best. In the same way, the use of other specific factors may be incentivated, for whatever reason (ethical, political, etc.). This is a matter of choice. For example, in emergy accounting, labour may sometimes have a very high emergy content and, as a consequences, certain labour-intensive processes could therefore be discouraged. To prevent this situation, policy makers could decide not to tax the labour-based part of total emergy. This would make labour more competitive with respect to other resources, but is again a matter of choice. Let us explore how Envitax can be implemented.

Envitax 1 The index to start with could be the transformity as indicated in Section 2.2. Resources, goods and services can be grouped into emergy classes,

---

They converge toward greater efficiency under conditions of higher prices but they do not return to higher resource intensity with a decrease in the price level.

characterised by orders of magnitude of transformities expressed as powers of 10 (for example, fossil fuels are of the order of 10^4 sej/J; electricity and biofuels, 10^6 sej/J; manufactured food and beverages, 10^7 sej/J; chemical fertilisers, 10^7 sej/J, etc.). These classes might be useful for taxation purposes. Taxation rates might be modulated according to the different classes of transformity. The higher the exponent, the higher the tax rate (Envitax 1), and consequently the cost of a given item. This would encourages consumer to use high transformity goods in appropriate and conservative way. Moreover, this would make inefficient processes less competitive and would stimulate competitive innovation. Agents would compete by choosing more efficient processes with lower transformity and, consequently, lower tax burden. Hence, Envitax 1 would have a double effect: on one hand, it would promote the use of more readily renewable resources and careful use of less renewable ones; on the other hand, it would promote more efficient use of all items in processes. Transformities, therefore, make it possible to select (and favour) items and processes that require lower direct and indirect support from the environment. This would favour efficiency of conversion.

Envitax 2 However, a low transformity only means that less emergy has been invested per unit of output. This is still an incomplete evaluation of the process. No information is provided about the mix of renewable and non-renewable sources contributing to it. As pointed out in Section 3.2, a low proportion of non-renewable inputs suggests an environmentally sound production system. If global sustainability (high economic yield and low environmental pressure) is our goal, the sustainability index (ESI), defined above, constitutes a further step ahead. Since it takes both economic (EYR) and ecological (ELR) sustainability into account, it is more comprehensive than transformity and probably more suitable for the aims of this paper. Its inverse (ELR/EYR) tends to zero the more a process is sustainable (Tables 1 and 2, item 14) and can be used as a Taxation Index (TI). A process with a lower TI should be taxed less, since for a given contribution to the economy, it causes less environmental stress. If two or more processes yielding the same product are considered, they should be taxed differently according to their TI (Envitax 2). The tax would therefore be a way to penalize processes which use technologies that are less compatible with the environment. By making these processes less competitive than more sustainable ones, it would incentivate technological innovation and research into processes with a lower TI. The ideal limit of this innovative process is a production process with an index equal to zero. Nothing prevents the policy maker from choosing a "floor" level in each class of processes, below which taxes would not be levied.

As stated, Envitax also involves using the tax revenue to restore the natural capital stressed by human activity. If we consider the environment as a fund instead of a stock, and if we agree that the fund must remain unchanged (in an economic sense, that is by restoring its ability to sustain a certain process), the idea of using tax revenue to restore natural capital is a step in the right direction. It would slow down depreciation of the fund and the cumulative effect on the biosphere of exploitation, making even greater, paraphrasing Georgescu-Roegen (1986), the lead time to wait for "Prometheus III", that is the technological innovation which will save us from disaster or which will allow us to recover from it. In a world characterised by imperfect knowledge, a prudent approach to irreversible problems is strongly required. To guarantee the dynamics of life processes, the environmental fund (the "agent" of these processes) must be kept in good condition. To this end, the goal of a sound environmental policy should be the preservation of the productive infrastructural functions of the biosphere. This is a dynamic concept that does not mean inalteration of the fund. Therefore, we suggest that the management principles used to run an enterprise or household should be used for environmental policy. According to these principles substantial activity has to be devoted to keeping capital items in a state as unchanged as possible.

Equipment cannot be maintained truly constant in time if the speed of deterioration is faster than the speed of maintenance.[9]

Standard economic theory has accustomed us to think in terms of stocks.

This has led us not only to deal with substitutability but also with substitutability between non-homogeneous elements, used for carrying out non-interchangeable tasks in the production process. This implicitly conveys the idea that scarcity is not a problem. On the contrary, if we think in terms of a fund, we see that the infrastructural productive function of the environment can be preserved only with appropriate process of maintenance. The fund suits the

---

[9]   According to Pearce et al. (1989), the notion that some suitably defined stock of capital should be kept constant is a crucial component of sustainability (Pezzey, 1989). A parallel can be drawn between an asset base yielding an income and natural capital. If "income" is defined to be the maximum real consumption expenditure that leaves society as "well-off" at the end of the period as at the beginning, it presupposes the deduction of expenditures to account for the depreciation or degradation of the asset base yielding that income. *"This implies to preserve some suitably defined capital stock with a view to ensuring that the constraints set does not tighten over time. It is, accordingly, income net of the expenditure needed to maintain such a suitably defined capital stock"* (Common and Perrings, 1992, p.9). As it can be noticed, the authors always use the word "stock", whereas, in our opinion, the concept of fund is much more appropriate.

dynamic view of the problem. In fact, it is both an input (entering the process with certain characteristics) and an output (emerging from the process with other characteristics). But, whereas we know the features of the entering fund, we are not always aware of the features of the outgoing fund.

Summing up, Envitax fosters an environmental policy that operates through: i) the supply side (driving processes towards more efficient use of resources); ii) the demand side (raising the price of goods) and iii) restoration of natural capital. A beneficial effect on environmental standards can be achieved in a shorter period than with the double dividend theory, because of this triple action.

With regard to the utilisation of tax revenues, it is worth noting that Envitax makes it possible to address the double-dividend issue in a new way. This issue is not addressed in a context of revenue neutrality, nor burdening environmental taxation with tasks it is not qualified for. It would be a by-product of the integrated approach to environmental problem. However, as for the carbon tax, results depend on many factors, especially wage rigidity.

Let us now investigate the possible effects. The new tax would push prices up, causing a drop in demand and negative effects on employment. However, investment of the tax revenues in restoration of the natural capital would counter these effects to some extent. Moreover, our proposal would enable a reduction in public spending which might have a positive effect on employment. Environmental expenditure is becoming an important part of the public budget. The earmarked environmental tax would at least bring about a saving in public resources, making tax reductions possible. This could lead to an increase in employment. This, however, is just a by-product of our proposal.

Envitax also avoids the problem mentioned at the end of Section 4. We noted that the beneficial effects of the tax on pollutant emissions could be neutralised by the increase in employment, due to the positive link between final demand and income. This does not seem to be the case in our model, because of the positive relation between tax revenues and restoring activity.

To conclude, two more characteristics of the new taxation scheme have to be considered. The first is prevalently economic: as stated, several studies indicate the regressive nature of taxes such as the carbon and energy tax, because they penalise prevalently low-income groups. The degree of regressiveness varies from country to country (Poterba, 1991; Pearson, 1993). Therefore, a tax proposed to correct a distortion in market prices would create distortion elsewhere. This is because some countries are heavy consumers of household energy, the demand of which is inelastic with respect to income. Envitax does not share this problem. It is paid on all kinds of items and services, and would penalise luxury goods, the demand of which is elastic. It is therefore less regressive than energy tax. Its quality dimension, which is a positive ecological property, turns out to be an economic advantage as well.

The second characteristic is economic and ecological. Emergy indices can be used to evaluate the local sustainability of a process. The relevance of this issue should not be overlooked. What is sustainable in a given environment could be unsustainable in another. In the long run, the local sustainability of a process becomes a competitive economic advantage, because the more the infrastructural production function of the environment is compromised, the greater the cost of producing a given item in the future.

Applying taxation schemes to case studies

In this Section we discuss data from different case studies to highlight the different way in which Envitaxes work compared with the carbon tax, proposed by the EU (Manne and Richels, 1993), and energy taxes, like Unitax, proposed by Slesser (1989).[10]

Biofuel production

Four different processes of biofuel production were investigated (i.e. ethanol from corn and from sugarcane, methanol from wood and biodiesel from rape seed). We calculated the energy demand and carbon dioxide release per unit energy produced, as well as emergy indicators, in order to assess the compatibility of the processes with environmental and economic constraints. The results are summarised in Table 1.

1) If energy demand per unit energy produced were the reference baseline (item 8 of Table 1), ethanol from sugarcane would be the most efficient process, and ethanol from corn the worst. Taxing energy demand would favour sugarcane-based systems and push ethanol producers to save fossil energy in the process. If fossil shortage were the main, or the only, concern, this would be an appropriate policy. Processes lacking the capacity to increase their energy efficiency would be highly penalised. But, an energy tax would strike earlier phases of each process, also discouraging processes that could contribute to subsequent phases of economic dynamics.

---

[10]   A critical survey of these taxation schemes can be found in Bimonte et al. (1996).

Table 1: **Comparison table of biophysical indicators for selected biofuel production systems**

		Methanol from Wood (#)	Biodiesel from Rapeseed (*)	Ethanol from Sugarcane (§)	Ethanol from Corn (**)
	Energy and Carbon Analysis:				
1	Yield (GJ/ha/yr)	15,80	35,70	74,40	50,90
2	Total energy demand (GJ/ha/yr)	14,40	28,70	42,30	96,30
3	Output/Input Energy Ratio (item 1/item 2)	1,10	1,24	1,76	0,53
4	CO_2 released (tons/ha/yr)	1,21	2,63	7,19	8,13
5	CO_2 avoided from substituting oil (tons/ha/yr)	1,70	3,58	8,02	5,49
6	CO_2 released/CO_2 avoided (item 4/item 5)	0,71	0,73	0,90	1,48
7	Net production of energy (GJ/ha/yr)	1,40	7,00	32,10	-45,40
8	Energy demand per unit energy delivered (J/J)	0,91	0,80	0,57	1,89
9	CO_2 released per unit energy delivered (g CO_2/MJ)	76,58	73,67	96,64	159,72
	Emergy Analysis:				
10	Transformity of fuel (sej/J)	1,35E+05	1,37E+05	2,63E+05	1,84E+05
11	Emergy yield ratio	2,28	2,16	3,74	1,45
12	Environmental Loading Ratio	1,00	2,32	11,40	4,96
13	Index of sustainability (item 11/item 12)	2,28	0,93	0,33	0,29
14	Taxation index based on item 13 (=1/item 13)	0,44	1,07	3,05	3,42

(*) Based on data from selected literature on Italian agricultural system (Ulgiati, 1995, unpublished Manuscript).

(§) Based on data from Farm System Lab, Dept of Food & Resource Economics, University of Florida Data Base (1985).

(#) Based on data from Ellington R.T., Meo M. and El-Sayed D.A., 1993. The net greenhouse warming forcing of methanol produced from biomass. Biomass and Bioenergy, 4(6), 405-418.

(**) Based on 1984 data for Italy (Triolo et al., 1984).

2) If CO_2 released per unit of energy produced were the reference baseline (item 9), sugarcane would no longer be the most efficient system, that is the one releasing the least carbon dioxide. In this specific case study, this aspect is due principally to the high organic matter content of Florida topsoil. Sugarcane production would increase soil oxidation and, consequently, a huge amount of CO_2 would be released by oxidised organic matter. Under this scenario, rape seed turns out to be the most efficient, producing the least carbon dioxide per MJ of energy yielded. If policy makers are concerned with greenhouse effect and global warming, they should tax CO_2 emissions. This policy would favour biodiesel and methanol producers.

3) If the concern were the global "entity" of the process, i.e. its global demand for environmental support, including free environmental inputs and indirect inputs embodied in materials, solar transformities (item 10) should be carefully considered. Methanol from wood has the lowest transformity (less solar emergy joules converging into the production of one joule of biofuel).

Taxing on the basis of transformity (Envitax 1) would give this process a comparative advantage and would probably force biodiesel and ethanol producers to increase the overall conversion efficiency of their processes to compete with methanol.

4) A different picture emerges from the values of the ESI and the related taxation parameter (items 13 and 14). From this point of view, methanol from wood is largely more sustainable than biodiesel and bioethanol, more than can be inferred from differences between the transformities. The higher the ESI (item 13), the higher the share of renewable emergy converging through the process. A taxation policy based on ESI would reward this higher "renewable efficiency" and environmental equilibrium (Envitax 2).

Agricultural production

Ethanol from corn yields no net energy. What meaning do emergy indices have in this case? Can the emergy approach be used to evaluate production process, other than energy conversion facilities?

Although ethanol production from corn is not a net energy source, its output can be used as a chemical input of industrial processes. The emergy indices enable comparison between processes that produce products other than fuels. In this case, nobody would care if the energy content of the agricultural product were lower than the energy input. In fact, food is not only a source of energy to an organism, it also provides special materials (proteins, oligoelements, etc), the value of which largely exceeds the organism's energy requirements. When energy is not the goal of a process, the main concern when comparing processes is not the output/input energy ratio, but to provide the same kind of product at the lowest relative "cost" per unit.

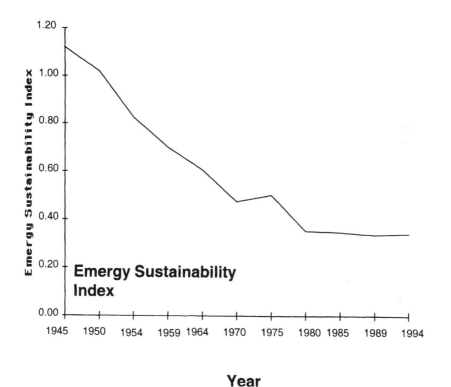

Year

Figure 3: **Change in the Emergy Sustainability Index of corn production in the US from 1945 to 1994 (Brown and Ulgiati, 1998, modif.)**

Figure 3 provides some last, but very important, information. Corn production in the US showed a steep decrease in ESI with time, until recently, when a stable situation seems to have been achieved. Industrial agricultural techniques presumably decreased the global sustainability of corn, until recently, when more appropriate technologies and concern for the environment stabilised system performance at an "optimum" level for available resources. Be this as it may, ESI index clearly reflects changes in performance over time.

Electricity production

Table 2 shows that electricity production by different kinds of power plants has different grades of environmental loading (Ulgiati, 1996). The ESI highlights differences between plants: the resulting taxation index is lowest for the wind power plant, and higher for hydroelectricity and geothermal production. If only traditional energy and carbon indicators were used, hydroelectricity would be taxed least because it produces less CO_2 and has a higher Energy Ratio. Traditional indicators do not count the value of the direct natural input of water.

Table 2: Comparison table of biophysical indicators for selected electricity power plants in Italy (Ulgiati, 1996)

		Wind (#)	Hydro (*)	Geothermal (§)
Energy and Carbon Analysis:				
1	Yield (GJ/yr)	13500	394000	328000
2	Total energy demand (GJ/yr)	1740	27300	128000
3	Output/Input Energy Ratio (item 1/item2)	7,76	14,43	2,56
4	CO2 released (tons/yr)	142	2230	10500
5	CO2 avoided from substituting oil (tons/ha/yr)	3390	99100	82500
6	CO2 released/CO2 avoided (item 4/item 5)	0,04	0,02	0,13
7	Net production of energy (GJ/yr)	11760	366700	200000
8	Energy demand per unit energy delivered (J/J)	0,13	0,07	0,39
9	CO2 released per unit energy delivered (g CO2/MJ)	10,52	5,66	32,01
Emergy Analysis:				
10	Transformity of electricity (sej/J)	6,14E+04	6,50E+04	1,85E+05
11	Emergy yield ratio	8,23	6,02	4,13
12	Environmental Loading Ratio	0,14	0,51	0,93
13	Emergy Sustainability Index (item 11/item 12)	58,79	11,80	4,44
14	Taxation index based on item 13 (=1/item 13)	0,02	0,08	0,23

(*) 85 MW, hydro-electric power plant sited at Castrocucco, Italy.

(#) 2.5 MW, wind power plant sited at Casone Romano, Italy.

(§) 20 MW, geothermal power plant sited at Castelnuovo V.C., Pisa, Italy.

197

ACKNOWLEDGEMENTS

The authors are indebted to Joseph Cornell (Syracuse University of New York), Mark T. Brown (University of Florida), Sandro Vercelli and Maurizio Franzini (University of Siena), for their precious comments on earlier versions of this paper.

Though the paper was written jointly by the authors, Sections 1 and 4 are mainly by Bimonte and Sections 3 and 6 mainly by Ulgiati.

REFERENCES

Barker, T., 1992. The carbon tax: economic and policy issues. Paper presented at the Workshop on "*Energy taxation and CO_2 emissions*". Fondazione ENI Enrico Mattei, Milan, Italy, March 19th, 1992.

Baumol, W.J.; Oates, W.E., 1988. *The theory of environmental policy.* 2nd ed., Cambridge, Cambridge University Press.

Bimonte, S., 1995. Pigouvian taxation in the case of undepletable externalities. *Note Economiche*, 1: 39-61.

Bimonte, S., Planeta, S. and Ulgiati, S., 1996. Environmental taxation for sustainable development. In: UlhØi, J.P. and Madsen H. (Editors), *Industry and the environment: praCtical applications of environmental management approaches in business.* The Aarhus School of Business (DK), 269-283.

Brown, M.T.; Ulgiati, S., 1997. Emergy-based indices and ratios to evaluate sustainability: monitoring economies and technology toward environmentally sound innovation. *Ecological Engineering* 9: 51-69.

Common, M. and Perrings, C., 1992. Towards an ecological economics of sustainability. *Ecological Economics*, 6: 7-34.

Ellington, R.T.; Meo, M.; El-Sayed D.A., 1993. The net greenhouse warming forcing of methanol produced from biomass. *Biomass and Bioenergy*, 4, (6): 405-418.

European Commission, 1993. *White paper on growth, competitiveness and employment*, Bruxelles.

Georgescu-Roegen, N., 1986. "Man and production". In: M. Baranzini, and R. Scazzieri, (Editors), *Foundations of economics: structures of inquiry and economic theory.* Basil Backwell, Oxford, pp. 247-280.

IPCC, Intergovernmental Panel on Climate Change, 1995. *Proceedings of the Eleventh Session*, Rome, 11-15 December 1995, in press.

Jones, R.A. and Ostroy, J.M., 1984. Flexibility and uncertainty. *Review of Economic Studies*. Jan.

Jørgensen S.E., 1997. *Integration of ecosystem theories: A pattern.* Kluwer Academic Publishers, Dordrecht (The Netherlands), Second Edition, 388 pp.

Lotka, A.J., 1922a. Contribution to the energetics of evolution. *Proceedings of the National Academy of Sciences*, U.S., 8: 147-150.

Lotka, A.J., 1922b. Natural selection as a physical principle. *Proceedings of the National Academy of Sciences*, 8: 151-155.

Majocchi, A., 1996. Green fiscal reform and employment: a survey, *Environmental and Resources Economics*, Vol. 8, no. 4, Dec.

Manne, A.S., Richels, G., 1993. The EC proposal for combining carbon and energy taxes: the implications for future CO_2 emissions. *Energy Policy*, 21, (1): 5-12.

Nordhaus, W.D., 1991. "Economic approach to greenhouse warming". In: R., Dornbusch and J.M., Poterba, (Editors), *Global warming: economic policy responses*. The MIT Press, Cambridge-Massachusetts, London-England, 33-66.

Oates, W. 1991. Pollution charges as a source of public revenues. *Working Paper no. 91-22*. Department of Economics, University of Maryland.

Odum, H.T., 1983. Maximum power and efficiency: a rebuttal. *Ecological Modelling*, 20: 71-82.

Odum, H.T., 1988. Self organization, transformity and information. *Science*, 242: 1132-39.

Odum, H.T., 1994. "The emergy of natural capital". In: A.M., Jansson, M., Hammer, C., Folke & R., Costanza, (Editors), *Investing in natural capital*. Island Press, Covelo, CA, pp. 200-12.

Odum, H.T., 1996. *Environmental accounting. Emergy and environmental decision making*. John Wiley and Sons, New York.

Odum, H.T.; Pinkerton, R.C., 1955. Time's speed regulator: the optimum efficiency for maximum power output in physical and biological systems. *American Scientist*, 43: 331-43.

OECD, 1991. *Environmental policy: how to apply economic instruments*. Paris.

OECD, 1993. *Environmental taxes in OECD countries: a survey*, Environmental monograph no. 71. Paris.

OECD, 1994. *The OECD jobs study: evidence and explanations*, Part II. Paris

OECD, 1997a. *Environmental taxes and green tax reform*. Paris.

OECD, 1997b. *Environmental policies and employment*. Paris.

Pearce, D., 1991. The role of carbon taxes in adjusting to global warming. *Economic Journal*. vol. 101: 938-948.

Pearce, D., Markandya, A. and Barbier, E.B., 1989. *Blueprint for a green economy*. Earthscan Publications, London.

Pearson, M. 1993. *Equity issues and carbon tax, in OECD, Climate change: designing a practical tax system*, Paris.

Pezzey, J., 1989. Economic analysis of sustainable growth and sustainable development, World Bank Environmental Department. *Working Paper n. 15*, World Bank, Washington D.C.

Piacentino, D., Smith, S., 1994. Taxes and subsidies in environmental fiscal reform. Paper presented at the International Workshop on *"Environmental taxation, revenues recycling and unemployment*", Fondazione ENI Enrico Mattei, Milan, Italy, Dec. 16th and 17th, 1994.

Poterba, J.M. (1991). "Tax policy to combat global warming: on designing a carbon tax", in R. Dornbush and J.M. Poterba (eds), *Global warming: Economic Policy Responses*, MIT Press.

Slesser, M., 1989. *Unitax: A new environmentally sensitive concept in taxation.* Ross-on-Wye, Hydatum Press.

Triolo, L., Mariani, A., Tomarchio, L., 1984. *L'uso dell'energia nella produzione agricola vegetale in Italia: bilanci energetici e considerazioni metodologiche.* ENEA, Italy, RT/FARE/84/12.

Ulgiati S., 1996. Efficiency and environmental sustainability indicators in the Italian electricity production. *Research Final Report*, Contract ENEA-University of Siena, December 1996.

Ulgiati, S. and Bastianoni, S., 1995. Monitoring a system's performance and efficiency by means of based indices and ratios, European Community, Environmental Research Programme. Area III, Economic and Social Aspects of the Environment. *Final Report*, Contract No.EV5V-CT92-0152.

Ulgiati, S.; Brown, M.T.; Bastianoni, S.; Marchettini, N., 1995. Emergy based indices and ratios to evaluate the sustainable use of resources. *Ecological Engineering,* 5: 519-531.

8. The Environmental Kuznets Curve:

A Critical Survey

Simone Borghesi[1]

Introduction

The relationship between economic growth and environmental quality has been the object of a large debate in the economic literature for many years.

This debate goes back to the controversy on the limits to growth at the end of the 1960s. At one extreme, environmentalists as well as the economists of the Club of Rome (Meadows et al. 1972) argued that the finiteness of environmental resources would prevent economic growth from continuing forever and urged a zero-growth or steady-state economy to avoid dramatic ecological scenarios in the future. At the other extreme, some economists (e.g. Beckerman 1992) claimed that technological progress and the substitutability of natural with man-made capital would reduce the dependence on natural resources and allow an everlasting growth path.

As Shafik (1994) has pointed out, in the past this debate lacked empirical evidence to support one argument or the other, remaining on a purely theoretical basis for a long time. This was mainly due to a lack of available environmental data for many years. However, it also reflected the difficulty of defining how to measure environmental quality. In the absence of a single criterion of environmental quality, several indicators of environmental degradation have been used to measure the impact of

---

[1] This work was previously published as working paper of the FEEM (Fondazione Eni Enrico Mattei). This study started when the author was in the Summer Internship Programme at the International Monetary Fund. The author would like to thank Jenny Ligthart, Valerie Reppelin and Liam Ebrill for helpful comments on the paper.

economic growth on the environment. However, different indicators yield different empirical results.

The World Development Report (1992) was one of the first studies to emphasise this issue. As shown in the Report (World Bank, 1992, Figure 4 p. 11), some indicators of environmental degradation (e.g. carbon dioxide emissions and municipal solid wastes) increase with income, which implies that they worsen with economic growth. Other indicators (such as the lack of safe water and urban sanitation) fall as income rises, indicating that - in these cases - growth can improve environmental quality. Finally, many indicators (e.g. sulphur dioxide and nitrous oxide emissions) show an inverted-U relationship with income, so that environmental degradation gets worse in the early stages of growth, but eventually reaches a peak and starts declining as income passes a threshold level (see Figure 1).

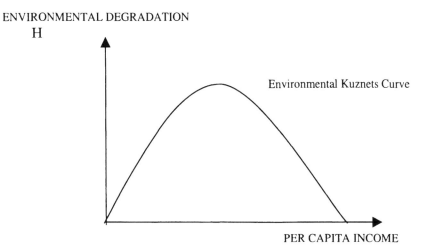

ENVIRONMENTAL DEGRADATION

Figure 1: Inverted-U (quadratic) curve

This inverted-U relationship has been defined as the Environmental Kuznets Curve (henceforth EKC) after Simon Kuznets, as it resembles the shape of the relationship that the Nobel Prize economist first observed between income distribution and economic growth.[2]

---

[2] It was probably Panayotou (1993) who first coined the term 'environmental Kuznets curve', although several contemporaneous studies observed this "bell-shaped" relationship between growth and environmental degradation in the early 1990s (Shafik 1994, Selden and Song 1994, Grossman 1995, Grossman and Krueger 1994). Notice that one can identify several versions of the EKC since there is not a universally accepted measure of

The object of the present paper is to provide a critical survey of the literature on the growth-environment relationship, focusing on the impact of growth on environmental quality. To the best of my knowledge, no one has yet attempted to give an overview of the many contributions that exist in this area, taking both early and recent studies into account.[3] In particular, the current review intends to determine whether and to what extent an EKC is empirically observed. In addition, attention will be focused on the policy implications of the empirical evidence. The main conclusion from the analysis of the literature is that the evidence on the environment-income relationship is not yet clear-cut and several methodological pitfalls cast doubts on the results that have been presented so far. Policy makers should therefore avoid simplistic recommendations based on current evidence. More specifically, the possibility that environmental degradation may eventually fall as income grows (as suggested by the alleged decreasing portion of the EKC) does not necessarily mean that growth will automatically solve the problems it causes in the early stages of development. Much work remains to be done to get a deeper understanding of the environment-income relationship. In this regard, the present paper emphasises the drawbacks of the cross-country studies that have been mainly used so far and the need to adopt a single-country approach, as suggested in some recent studies.

The structure of the paper is as follows. Section II investigates the effects of growth on environmental quality to establish the theoretical underpinnings of the EKC. Section III is divided in two parts. The first examines the empirical evidence on the EKC that emerges from the cross-country studies to determine: 1) for which environmental indicators such a curve exists; 2) at what income level environmental degradation starts to decrease decreasing. The second part explores how the evidence changes when we follow the evolution of the environment-income relationship in a single country over time rather than inferring it from cross-country analyses. Section IV draws attention to limits of current studies (both cross- and single-country) that restrict the reliability of the evidence in favour of the EKC.

---

environmental degradation (see section IV.D for a more detailed discussion of this aspect).

[3] Previous reviews have focused either on the early studies (Pearson 1994) or on part of the later contributions (Barbier 1997). Moreover, the literature review in Pearson is organized by studies, which is made possible by the small number of works that he examines. Given the difference in the results across the environmental indices, the present survey is organized by indicator, thus providing a different perspective with respect to former essays.

Section V discusses the policy implications emerging from the literature on the EKC, especially for the developing countries that are now on the upward part of the alleged curve. Some concluding remarks will follow.

Conceptual background of the EKC: effects of growth on the environment

As Grossman (1995) first suggested, it is possible to distinguish three main channels whereby income growth affects the quality of the environment. In the first place, growth exhibits a scale effect on the environment: a larger scale of economic activity leads per se to increased environmental degradation. This occurs because increasing output requires that more inputs and thus more natural resources are used up in the production process. In addition, more output also implies increased wastes and emissions as by-products of the economic activity, which contributes to worsen the environmental quality. In the second place, income growth can have a positive impact on the environment through a composition effect: as income grows, the structure of the economy tends to change, gradually increasing the share of cleaner activities in the Gross Domestic Product. In fact, as Panayotou (1993, p. 14) has pointed out, environmental degradation tends to increase as the structure of the economy changes from rural to urban, from agricultural to industrial, but it starts falling with the second structural change from energy-intensive heavy industry to services and technology-intensive industry.

Finally, technological progress often occurs with economic growth since a wealthier country can afford to spend more on research and development.[4]

This generally leads to the substitution of obsolete and dirty technologies with cleaner ones, which also improves the quality of the environment. This is known as the technique effect of growth on the environment.

An inverted-U relationship between environmental degradation and per capita income suggests that the negative impact on the environment of the scale effect tends to prevail in the initial stages of growth, but that it

---

[4] For instance, Komen at al. (1997) examine data on 19 OECD countries between 1980 and 1994 and show that the income elasticity of public research and development expenditures for environmental protection is approximately equal to one. Notice that technological progress can be seen as both the cause and effect of economic growth.

will eventually be outweighed by the positive impact of the composition and technique effects that tend to lower the emission level. The income elasticity of environmental demand is often invoked in the literature as the main reason to explain this process. As income grows, people achieve a higher living standard and care more for the quality of the environment they live in. The demand for a better environment as income grows induces structural changes in the economy that tend to reduce environmental degradation. On one hand, increased environmental awareness and "greener" consumer demand contribute to shift production and technologies toward more environmental-friendly activities. On the other hand, they can induce the implementation of enhanced environmental policies by the government (such as stricter ecological regulations, better enforcement of existing policies and increased environmental expenditure). This also contributes to shift the economy towards less polluting sectors and technologies. Hence, the demand for a better environment and the resulting policy response are the main theoretical underpinnings behind the decreasing path of the EKC (Grossman, 1995 p.43).[5]

Another argument has been advanced in the literature to explain the bell-shaped environment-income pattern. It has been suggested (World Bank 1992, Unruh and Moomaw 1998) that the existence of an endogenous self-regulatory market mechanism for those natural resources that are traded in markets might prevent environmental degradation from continuing to grow with income. In fact, early stages of growth are often associated with heavy exploitation of natural resources due to the relative importance of the agricultural sector. This tends to reduce the stock of natural capital over time.

The consequent increase in the price of natural resources reduces their exploitation at later stages of growth as well as the environmental degradation associated with it. Moreover, higher prices of natural resources also contribute to accelerate the shift toward less resource-intensive

---

[5] Many authors have claimed that the environment is a luxury good, that is, environmental demand does not simply increase as people get richer, but grows faster than income. However, two recent contributions have challenged this interpretation. McConnell (1997) has proved that the assumption that the environment is a luxury good is neither a necessary nor a sufficient condition to obtain an EKC. In fact, he shows that pollution can decrease even if the demand for environmental quality is inelastic with respect to income. In the same way, under specific conditions, pollution may increase even if the demand for the environment is very elastic. Kristrom and Riera (1996) went even further, questioning the assumption that the environment is a luxury good. They estimated the income elasticity of environmental improvements in several European countries and found that in many cases this elasticity is less than one.

technologies (Torras and Boyce, 1998).[6] Hence, not only induced policy interventions, but also market signals can explain the alleged shape of the EKC.

Empirical evidence on the environment-income relationship

The above discussion indicates the conceptual arguments that make the EKC conceivable from a theoretical viewpoint. We now ask whether empirical evidence really supports this pattern and what indicators follow it? Given the lack of long time-series of environmental data, most empirical studies have adopted a cross-country approach to address this question. The present section examines the results and main limitations of these studies and indicates the single-country approach as an alternative method for future research.

Cross-country studies

All the studies on the EKC address the following common questions:

1. Is there an inverted-U relationship between income and environmental degradation?
2. If so, at what income level does environmental degradation start declining?

As we shall see, both questions have ambiguous answers.

In the absence of a single environmental indicator, the estimated shape of the environment-income relationship and its possible turning point generally depend on the index considered. In this regard, it is possible to distinguish three main categories of environmental indicators that have been used in the literature: air quality, water quality and other environmental quality indicators.

As to air quality indicators, there is strong, but not overwhelming evidence of an EKC. A distinction is conventionally made in the literature between local and global air pollutants (e.g. Grossman 1995, Barbier 1997).[7]

---

[6] As Unruh and Moomaw (1998) point out, the increase in the oil price that occurred during the 1970s promoted the shift to alternative sources of electric power production.

[7] Grossman (1995) was among the first to draw the distinction between local and global

The measures of urban and local air quality (sulphur dioxide, suspended particulate matters, carbon monoxide and nitrous oxides) generally show an inverted-U relationship with income. This outcome, that emerged in all early studies, seems to be confirmed by more recent works (Cole et al., 1997).

However, there are major differences across indicators as to the turning point of the curve: carbon monoxide and especially nitrous oxides show much higher turning points than sulphur dioxide and suspended particulate matters (see Table 1). Moreover, there are also large differences across studies that focus on the same indicator. For instance, Selden and Song (1994) estimate a turning point for suspended particulate matters three times higher than that found by Shafik (1994). Similar large differences occur in the case of sulphur dioxide (see Table 1).[8]

When emissions of air pollutants have little direct impact on the population the literature generally finds no evidence of an EKC. In particular, both early and recent studies find that emissions of global pollutants (such as carbon dioxide (CO_2)) either monotonically increase with income or start declining at income levels well beyond the observed range. Moreover, Cole et al. (1997) have recently pointed out that even in studies that find a peak (however high) in the CO_2 curve, the alleged turning point has a very large standard error. This implies that estimates of the CO_2 turning point are quite unreliable, casting doubts on the possible downturn of the CO_2 curve.

For water quality indicators, empirical evidence of an EKC is even more mixed. However, when a bell-shaped curve does exist, the turning point for water pollutants is generally higher than for air pollutants. Three main categories of indicators are used as measures of water quality: 1. concentration of pathogens in the water (indirectly measured by faecal and total coliforms), 2. amount of heavy metals and toxic chemicals discharged in the water by human activities (lead, cadmium, mercury, arsenic and nickel) and 3. measures of deterioration of the water oxygen regime

---

air pollutants, which is often adopted also in recent contributions. However, this distinction is not clear-cut: some local pollutants (e.g. sulfur dioxide (SO_2)), may travel for hundreds of miles, so they can be considered both local and global air quality indicators.

[8] The differences in these results can be explained by differences in the way pollution is measured as well as in sample size. For instance, Selden and Song (1994) measure the *flow* of emissions of local air quality indicators in 22 countries, whereas Shafik (1994) focus attention on the *stock* of the same indicators using a much larger database (up to 149 countries).

(dissolved oxygen, biological and/or chemical oxygen demand).[9] As Table 1 shows, there is evidence of an EKC for some indicators (especially in the latter category), but many studies reach conflicting results as to the shape and peak of the curve.[10] Several authors (Grossman and Krueger 1994, Shafik 1994, Grossman 1995) find evidence of an N-shaped curve for some indicators: as income grows water pollution first increases, then decreases and finally rises again (Figure 2).

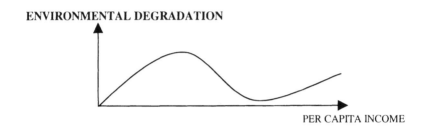

ENVIRONMENTAL DEGRADATION

PER CAPITA INCOME

Figure 2: N-shaped (cubic) curve

Thus, the inverted-U curve might correspond just to the first two portions of this more complex pattern. The existence of an N-shaped curve seems to imply that at very high income levels; the scale of the economic activity becomes so large that its negative impact on the environment cannot be counterbalanced by the positive impact of the composition and technology effects mentioned above.[11]

Finally, in the absence of a single definitive measure of environmental quality, many other environmental indicators have been used to test the EKC hypothesis. In general, for most of these indicators there seems to be little or no evidence of a Kuznets-type story. Both early and recent studies (Shafik 1994, Cole et al. 1998) find that environmental problems having direct impact on the population (such as access to urban sanitation and clean water) tend to improve steadily with growth. On the contrary, when

---

[9] See Grossman (1995) for a detailed description of the environmental problems and health risks caused by each pollutant.

[10] Compare, for instance, the results obtained by Grossman (1995) and Grossman and Krueger (1994) with those achieved by Shafik (1994) (Table 1).

[11] Shafik (1994, p.765) has advanced the hypothesis that the increase in rivers pollution at high-income levels typical of an N-shaped curve might occur because "people no longer depend directly on rivers for water and therefore may be less concerned about river water quality".

environmental problems can be externalised (as in the case of municipal solid wastes) the curve does not even fall at high income levels. As to deforestation, the empirical evidence is controversial.[12] Some studies find an inverted-U curve for deforestation with the peak at relatively low income levels (e.g. Panayotou 1993), whereas others conclude that "per capita income appears to have little bearing on the rate of deforestation" (Shafik 1994, p.761). Finally, even when an EKC seems to apply (as in the case of traffic volume and energy use), the relative turning points are far beyond the observed income range.

Summing up, three main stylised facts that provide the answer to our initial questions seem to emerge from cross-country studies: 1. only some indicators (mainly air quality measures) follow an EKC; 2. an EKC is more likely for pollutants with direct impact on the population rather than when their effects can be externalised;[13] 3. in all cases in which an EKC is empirically observed, there is still no agreement in the literature on the income level at which environmental degradation starts decreasing.

From cross-country to single-country studies

As shown above, cross-country studies suggest that the EKC may only be a valid description of the environment-income relationship for a subset of all possible indicators. However, Roberts and Grimes (1997) have recently questioned the existence of an EKC, even for indicators that seem to follow this pattern. They observe that the relationship between per capita GDP and carbon intensity changed from linear in 1965 to an inverted-U in 1990.[14]

How can we explain the modification in the curve shape over the last thirty years? Roberts and Grimes (1997, p.196) argue that the Kuznets-type curve that we observe for carbon intensity today is the result of environmental improvement in developed countries in these last decades

---

[12] As Panayotou (1993) pointed out, the rate of deforestation is particularly important as a measure of environmental degradation for two reasons. Firstly, it can be taken as a proxy variable for the depletion of natural resources. Secondly, together with land use changes, deforestation accounts for about 17-23% of total anthropogenic carbon dioxide emissions (World Resource Institute 1996).

[13] This seems to reflect the existence of a free-rider problem. In fact, as Shafik (1994, p.770) argues, "where environmental problems can be externalized, there are few incentives to incur the substantial abatement costs associated."

[14] Carbon intensity is defined as carbon dioxide emissions per unit GDP. See section IV.D for discussion on the choice of the measure of environmental degradation.

and "not of individual countries passing through stages of development." In fact, the data set shows that carbon intensity fell steadily among high income countries in the period 1965-90, but increased among middle- and low-income nations, with a marked increment in the latter group. Therefore, the EKC that emerges in the cross-section analysis "may simply reflect the juxtaposition of a positive relationship between pollution and income in developing countries with a fundamentally different, negative one in developed countries, not a single relationship that applies to both categories of countries" (Vincent 1997, p. 417). For this reason, Vincent (1997) claims that the cross-country version of the EKC is just a statistical artefact and should be abandoned. In fact, as Stern et al. (1994) have argued, "more could be learnt from examining the experiences of individual countries at varying levels of development as they develop over time." These considerations have given rise to a new line of research based on single-country analysis. This econometric approach achieves some surprising results that cast serious doubts on the reliability of the indications emerging from cross-country studies.

Vincent (1997) examines the link between per capita income and a number of air and water pollutants in Malaysia from the late 1970s to the early 1990s. Two main conclusions emerge from this single-country study. First, cross-country analysis may fail to predict the income-environment relationship in single countries, as it occurs in the case of Malaysia. Second, none of the pollutants examined by Vincent shows an inverted-U relationship with income. Contrary to cross-section analysis, several measures indicate that increments in the income level may actually worsen environmental quality. It can be argued that the results achieved by Vincent hinge heavily on specific features of the country in question and cannot be extended to other countries. However, de Bruyn et al. (1998) reach similar conclusions following other individual countries over time. They investigate emissions of several air pollutants (sulphur dioxide, carbon dioxide and nitrous oxides) in four OECD countries (Netherlands, West Germany, UK and USA) between 1960 and 1993 and find them to be positively correlated with growth in almost every case.[15] However, these

---

[15] The only exception (out of the 12 cases that they observe) is sulfur dioxide emissions that decreased monotonically with per capita income in the Netherlands. In general, the growth parameter (which is taken by the authors as a measure of the size effect) is estimated to be around 1, so that – ceteris paribus – income and emissions tend to grow at the same speed. The impact of growth on emissions can be counteracted by the reduction of emissions due to technological and structural factors (i.e. the composition and technique effects mentioned above). However, the authors find that in some cases these effects turn out to be statistically insignificant, which explains why the size effect tends to prevail.

conclusions are questioned by Carson et al. (1997) who find the opposite result in a single-country study on the Unites States. Using data collected by the Environmental Protection Agency from the 50 US states, Carson et al. (1997) find that per capita emissions of air toxics decrease as per capita income increases.

In conclusion, all current single-country studies seem to suggest that the EKC need not hold for individual countries over time. However, different studies reach conflicting results as to the effects of growth on the environment. Therefore, further research is needed to understand the evolution of environmental degradation relative to income in a single country over time. In particular, both Vincent (1997) and Carson et al. (1997) are cross-regional studies; therefore they are also subject to the critiques to the cross-country approach mentioned above. In fact, cross-country studies implicitly assume that all countries will follow the same pattern in order to infer the environment-income relationship of a single country over time. As mentioned above, this assumption does not seem to be supported by empirical evidence.

Similarly, in order to infer the environmental degradation of the whole country over time, cross-regional studies implicitly assume that all regions in a given nation will follow the same pattern. For some countries, however, regional differences can be very significant. Thus, the environment-income relationship may not only differ across nations, but also across regions of the same country.[16] Hence, although current single-country studies tend to go in the right direction, a time-series approach seems more appropriate than a cross-regional one to examine individual countries over time and this is the line of research that single-country analyses should develop in the future.

Limitations of current studies

As many authors have underlined (e.g. Grossman and Krueger 1994), knowing the shape of the environment-income relationship could help policy makers to formulate appropriate environmental policy. However, current results do not seem completely reliable for this purpose. We already mentioned why cross-sectional studies (both cross-country and cross-region) limit the validity of the evidence at disposal. In this section, we look at some other drawbacks of the current literature that should

---

[16] However, differences across regions are generally smaller than those across countries.

induce us to use the available results with particular caution for policy aims.

Data problems

The first and most obvious limitation of the studies on the EKC is the lack of good data on environmental indicators.[17] Even when such data is available, it appears to be unreliable in some low-income countries because of data collection problems. Moreover, the existence of definitional differences across countries raises problems of data comparability, casting serious doubts on the cross-country approach (Shafik 1994, Carson et al. 1997).

One important consequence of the lack of data is that many studies use estimates rather than actual measures of environmental indicators (see Table 1). Such estimates are based on rates of conversion from economic data "both of which can be unreliable, especially in developing countries" (Kaufmann et al., 1998). In some cases (e.g. carbon dioxide) the estimates are computed by applying emission coefficients to national consumption of various kinds of fuel. In other cases (e.g. sulphur dioxide and other air pollutants) they are calculated by multiplying this national consumption "by coefficients that reflect the contemporaneous abatement practices in each country" (Grossman 1995, p.24).

Beyond data quality and comparability, current studies may also suffer from sample selection bias. In fact, monitoring stations that collect data on pollution are often situated where pollution is potentially more severe. Thus, for instance, most stations are in towns or along rivers suspected of high pollution. Therefore, the results are likely to reflect local conditions and, in some circumstances, pollution might be overestimated. On the other hand, most of the available data is from developed countries. However, a large contribution to global pollution comes from many developing countries for which data is not available. Hence, the sample selection made in cross-country studies may underestimate the level of pollution.

Reduced-form models

Both cross- and single-country studies are based on reduced form models.[18]

---

[17] In general, environmental data is much scarcer than economic statistics. Even in OECD countries that have long time series, environmental indicators are only available from the 1970s.

[18] As it is well known, this means that the current endogenous variable (environmental quality) is expressed only as a function of predetermined variables.

As de Bruyn et al. (1998) point out, these models enable economists to estimate the influence of income on environmental quality. However, they give no indication about the direction of causality, namely whether growth affects the environment or the other way around. In other words, reduced-form relationships "reflect correlation rather than a causal mechanism" (Cole et al. 1997, p.401). In reality, environmental quality is likely to have a feedback effect on income growth (Stern et al. 1994, Pearson 1994). As a matter of fact, the environment is a major factor of production in many underdeveloped countries that heavily rely on natural resources as a source of output.

Therefore, environmental degradation in these countries is likely to reduce their capacity to produce and hence to grow. Moreover, several studies point out that high pollution levels may reduce worker productivity and thus economic output. Hence, a simultaneous-equation model may be more appropriate for understanding the environment-income relationship.[19]

Limitations of econometric techniques

Besides the problems mentioned so far, there are also other limitations to the validity of current EKC studies (both cross- and single-country). One of the main criticisms concerns the choice of specific functional forms to estimate the environment-income relationship. Most of the literature has examined reduced-forms in which the environmental indicator is a quadratic or cubic function of income. However, neither the quadratic nor the cubic function can be considered a realistic representation of the environment-income relationship. As Cole et al. (1997) pointed out, a cubic function implies that environmental degradation will eventually tend to plus or minus infinity as income grows over time. Similarly, a quadratic concave function implies that environmental degradation could eventually tend to zero (or even become negative) at sufficiently high income levels, which is not supported by empirical evidence.[20] Another drawback of the quadratic function is that it is symmetrical, that is, the uphill portion of the curve has the same slope as the downhill part. This implies that,

---

[19] A simultaneous-equation model is a system of equations in which environmental quality and income are both endogenous variables. To the best of my knowledge, there has only been one attempt (Dean 1996) to use this approach so far. However, Dean applies this method to investigate the impact of trade liberalization on environmental quality in developing countries, which goes beyond the scope of the present paper.

[20] In fact, there is no evidence that any country has environmental degradation close to zero.

when income goes beyond some threshold level, environmental degradation will decrease at the same rate as it previously increased. This is also very unlikely, as many forms of environmental degradation can be extremely difficult to undo. For instance, most pollutants tend to accumulate and persist for a long time, so that they are generally much harder to mitigate than to produce. Hence, as Pearson (1994, p.212) argues, more sophisticated techniques of curve fitting should be investigated in the future so that our findings are not determined by the specific functional form chosen.

The use of unrefined econometric techniques concerns not only the choice of regression models, but also the estimation method. This is another reason suggesting a cautious attitude to the empirical evidence of some studies. For instance, most of the early studies used Ordinary Least Square (OLS) estimations without correcting for heteroscedasticity and autocorrelation of the residuals. However, Carson et al. (1997) point out that the variance of the error terms may differ across countries or regions.[21] The residuals are also likely to be autocorrelated because of common shocks (e.g. the oil shock) that affect several countries simultaneously (Unruh and Moomaw 1998). In all these cases, OLS estimates of the standard errors turn out to be biased. However, this weakness mainly concerns the early studies and has been generally corrected in recent contributions by using Generalised-Least Square (GLS) estimates.

Choice of the scaling factor of environmental degradation

Another problem that arises in the empirical literature is the choice of the scaling factor to be used in the regression model. While all studies agree on using per capita GDP as the independent variable on the horizontal axis, one can distinguish three main variants in the literature for the dependent variable:

1. per capita emissions
2. total emissions
3. emission intensity (i.e. per unit of GDP).

---

[21] For instance, Carson et al. (1997) find strong evidence of heteroscedasticity for air emissions across the US states, the variance of the residuals being a decreasing function of income.

These measures can have very different implications. This is evident if we look at a potentially different shape of the EKC. As Common (1995) noted, the Kuznets-type pattern with pollution that first increases and then decreases with income is consistent with two possible cases: (a) at sufficiently high income levels, the quadratic curve falls to zero (Figure 1), (b) at sufficiently high income levels, the curve tends to a lower bound k (Figure 3).[22]

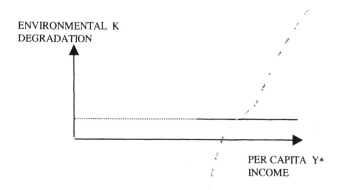

ENVIRONMENTAL K
DEGRADATION

PER CAPITA Y*
INCOME

Figure 3: Environmental Kuznets Curve with lower bound
Source: Common 1995

We have already discussed case (a). As to case (b), if the vertical axis measures total emissions the shape of the curve implies that emissions will become constant at a sufficiently high-income level Y*. However, if we measure emission intensity on the vertical axis, the existence of a lower bound implies that total emissions will not be constant, but will grow at the same rate as income so that emissions will tend to infinity in the long run.

In addition, each version of the EKC sheds light on aspects that do not emerge in the other two variants. For instance, the scatter diagram for cross-country CO_2 emission intensity in 1995 (Figure 4) reveals extremely high values of this variable in former Soviet Union countries.[23]

---

[22] Common (1995) argues that case (b) is more interesting than (a) as it avoids the unrealistic implication that pollution eventually goes to zero or becomes negative.

[23] This occurs because former Soviet Union countries have both high CO_2 emissions and low incomes. The acronyms used in Figure 4 are as follows: UKR = Ukraine, AZR = Azerbaijan, KZK = Kazakhstan, UZB = Uzbekistan, RUS = Russia.

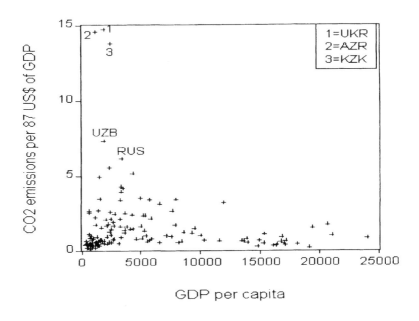

Figure 4: 1995 CO2 emission intensity

On the contrary, the pollution impact of these regions does not emerge if we look at cross-country per capita emissions of CO_2 in the same year (Figure 5).[24]

---

[24] The following acronyms have been used in Figure 5: LUX = Luxembourg, ARE = United Arab EmiratR = Bahrain, SGP = Singapore, CHE = Switzerland.

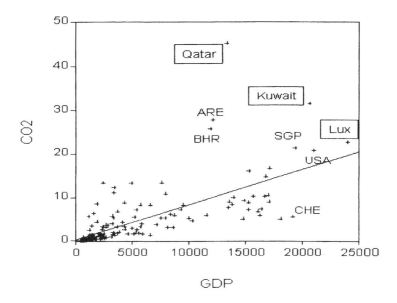

Figure 5: Cross country per capita emissions

In this case, the outliers are mainly the oil producing countries that have high emissions and low population levels.

In general, the correct choice of EKC version should depend on the environmental indicator considered. For instance, the EKC in terms of per capita emissions is probably more correct than the other two versions when the main source of environmental depreciation is overexploitation of natural resources caused by population growth, whereas the emission intensity version provides a deeper insight when pollution is due mainly to heavy industry.

Some studies (Shafik 1994, Kaufman et al. 1998) have proposed pollutant concentration as an alternative indicator of environmental degradation. This is probably the most appropriate indicator when one examines global pollutants since their stock contributes to global warming more than their emissions (the so-called "stock externality" problem). This casts further doubts on the evidence in favour of the EKC. In fact, a convex relationship often emerges in studies that measure concentration rather than emissions of global pollutants (Kaufman et al. 1998 for SO_2, Shafik 1994 for CO_2).

Policy implications

The shape of the environment-income relationship has critical policy implications. The alleged form of the EKC has lead some authors to conclude that current environmental degradation might be only a temporary phenomenon and that it is possible to "grow out" of the environmental problems in the long run (Beckerman 1992). If so, policy-makers should promote faster growth rates to overcome the income turning point as soon as possible. However, even if we neglect the flaws of the empirical studies and accept the EKC as a stylised fact for the sake of the argument, there are several reasons to question this conclusion.

As Panayotou (1993) has underlined, a policy that devotes most resources to growth is not necessarily an optimal one. In fact, achieving the downturn of the EKC may be a very long process that takes decades, the more so the longer one waits to intervene.[25] In fact, emissions and the consequent environmental degradation often tend to accumulate over time. Therefore, delaying intervention to later stages of growth may result in prohibitively high abatement costs. If so, environmental damage that is physically reversible could become economically irreversible. In addition, the literature has largely been concerned with the income level at which the turning point occurs. However, the height of the curve may be even more important. If emissions or concentrations at the vertex of the parabola are above some threshold level, we may enter that "shadow area" where the damage is unknown and potentially irreversible (Figure 6).

---

[25] Selden and Song (1994) estimate that global emissions of all air pollutants that show an EKC in the cross-section analysis will keep on growing in future decades. This is what one would expect, since countries now on the upward portion of the curve often have the fastest rates of economic and population growth. Therefore, "emissions will not return to current levels before the end of the next century unless concerted actions are taken" (Selden and Song 1994, p.161).

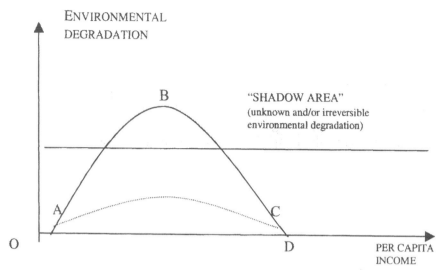

Figure 6: "Tunnelling through" the EKC

This implies that environmental degradation may become irreversible before we reach the top of the curve. If so, it might be impossible to exploit the decreasing path of the EKC at a future date. This possibility should not be neglected, especially because empirical evidence suggests that the EKC is not stable, but tends to shift and change in shape with time (Roberts and Grimes 1997).

For all these reasons, a policy of "wait and see" based on acritical faith in the EKC may have vast negative effects on the environment in the future.

On the contrary, we should intervene to "tunnel through" the curve (Munasinghe 1998), building a bridge between the upward and downward portions of the EKC, without letting environmental problems reach their peak level. As Panayotou (1993) has argued, several policies can be implemented to flatten out the curve. For instance, eliminating policy distortions (e.g. energy and agrochemical subsidies) or enforcing property rights over natural resources may both serve this purpose.[26]

These considerations are particularly important for developing countries currently on the upward part of the curve. There is good reason to believe that these countries may not be able to follow the same path as developed countries in the past. In the first place, as Unruh and Moomaw (1998, p.222) have claimed:

---

[26] See Panayotou (1993) for a thorough discussion of these environmental policies.

"...it is not certain whether 'stages of economic growth' is a deterministic process that all countries must pass through, or a description of the development history of a specific group of countries in the 19th and 20th centuries that may or may not be repeated in the future".

In the second place, the environmental conditions in which the South is developing today are much different from the ones faced by the North in the past. In fact, the stock of greenhouse gases inherited by today's developing countries is certainly higher than that met by the developed countries in the early stages of their development. As the so-called "stock externality" issue suggests, it is this stock, rather than the current flow of emissions, that contributes most to global warming and the damage that this creates. Hence, if we could measure actual environmental degradation rather than emissions on the vertical axis, the EKC of the newly developing countries might shift upward with respect to the EKC of the industrialised ones for a given income level.

Finally, Roberts and Grimes (1997) indicate another reason why the South may be unable to take the path followed by the North. Some of the environmental improvements in the North were made possible by relocating its most polluting, energy-intensive industries in the South (Hettige, Lucas and Wheeler 1992). However, the South will be unable to find in turn some other countries where these industries can be shifted in the future. Moreover, even if we transferred the least polluting and most energy-saving technologies from North to South, this might not necessarily improve the environmental quality in the latter unless other socio-economic reforms are undertaken.

Quoting Roberts and Grimes (1997, p.196):

> even identical industries operating in non-wealthy countries face obstacles making them less efficient in energy and carbon terms, such as poor roads, inefficient energy sources and local shortages of well-educated high-tech workers.

These considerations call for an international environmental policy that is different from the one recently developed in the Kyoto agreements. The North now has the whole burden of cutting emissions, while the South has been left free to pollute. This policy reflects the belief (partially nourished by a misinterpretation of the EKC) that the developing countries first need to grow which will automatically lead them to address their environmental problems in the future. However, increasing pollution in the developing countries may have adverse effects on developed nations. As a matter of fact, issues such as global warming affect all countries irrespectively of the nation where emissions occur: one unit of pollutant contributes equally to the greenhouse effect wherever it is emitted. Therefore, if negative externalities from the South to the North are strong enough, the curve of the environmental damage due to pollution could rise again in the wealthiest countries.

As stated by Roberts and Grimes (1997), sustainability should be addressed at all levels of development if we are to avoid this risk. This does not mean introducing the North's high environmental standard also in the South from the beginning, but ensuring that environmental interventions accompany the financing policies of the development assistance agencies in the South. This is particularly important if we do not want developing countries to simply mimic the past experience of industrialised nations, but rather to learn from it.

Conclusions

In the last few years, there has been renewed interest in the relationship between income growth and environmental quality. A remarkable number of new contributions have investigated this relationship empirically, correcting for some of the drawbacks of early studies. Despite the use of more sophisticated econometric techniques, there is still no clear-cut evidence to support the existence of the EKC. As shown by this review of the empirical evidence, there is hardly ever agreement among different studies on the many environmental indicators examined in the literature. The lack of consensus concerns not only the turning point of the EKC, but also its very existence.

Moreover, even when an inverted-U relationship does appear, it may be an artificial result of the cross-country approach. This approach seems inadequate to predict the future evolution of the environment-income relationship: industrialised countries may have moved along an inverted-U pattern in the past, but this does not imply that developing countries will or should follow the same pattern today. Therefore, future research should use time-series analysis to determine the pollution trajectories of each country over time, improving on the lines indicated by recent single-country studies.

This is particularly important for developing countries, many of which are in tropical areas where the fauna and flora have generally low resilience.

A misdirected growth policy based on acritical faith in the EKC could have large and potentially irreversible effects in these nations, ruling out the possibility to run along the decreasing part of the curve in the future.

REFERENCES

Barbier, E., 1997, "Introduction to the environmental Kuznets curve special issue", Environment and Development Economics, Vol. 2, pp. 369-381, Cambridge University Press.

Beckerman, W., 1992, "Economic growth and the environment: whose growth? Whose environment?", World Development, Vol. 20, n. 4, pp. 481-496.

Carson, R.T., Jeon, Y., and McCubbin, D. R. 1997, "The relationship between air pollution emissions and income: US data", Environment and Development Economics, Vol. 2, pp. 433-450, Cambridge University Press.

Cole, M.A., Rayner, A.J., and Bates, J.M., 1997, "The environmental Kuznets curve: an empirical analysis", Environment and Development Economics, Vol. 2, pp. 401-416, Cambridge University Press.

Common, M.S., 1995, "Sustainability and policy", Cambridge University Press, Cambridge, UK.

Dean, J.M., 1996, "Testing the impact of trade liberalization of the environment", Johns Hopkins University, Washington D.C., mimeo.

de Bruyn, S.M., van den Bergh, J., Opschoor, J.B., 1998, "Economic growth and emissions: reconsidering the empirical basis of environmental Kuznets curve", Ecological Economics, Vol. 25, pp. 161-175.

Grossman, G.M., 1995, "Pollution and growth: what do we know?", in "The economics of sustainable development" edited by Goldin I. and Winters L.A., Cambridge University Press, pp. 19-45.

Grossman, G.M., and Krueger, A.B.,1994, "Economic growth and the environment", NBER Working Paper n. 4634, February; also in Quarterly Journal of Economics Vol. 110 (1995), pp.353-377

Hettige H., Lucas R., and Wheeler D. 1992, "The toxic intensity of industrial pollution: global patterns, trends and trade policy", American Economic Review 82(2), pp. 478-481.

Holtz-Eakin D., Selden T.M. (1995) "Stoking the fires? CO_2 emissions and economic growth", Journal of Public Economics, Vol. 57, pp. 85-101.

Kaufmann, R.K., Davidsdottir, B., Garnham, S., and Pauly, P., 1998, "The determinants of atmospheric SO_2 concentrations: reconsidering the environmental Kuznets curve", Ecological Economics, Vol. 25, pp. 209-220.

Komen, M., Gerking, S., and Folmer, H., 1997, "Income and environmental R&D: empirical evidence from OECD countries", Environment and Development Economics, Vol. 2, pp. 505-515, Cambridge University Press.

Meadows, D.H., Meadows, D.L., Randers J., and Behrens, W., 1972, "The limits to growth", Universe Books, New York, USA.

Munasinghe, M., 1998, "Is environmental degradation an inevitable consequence of economic growth: tunnelling through the environmental Kuznets curve", Ecological Economics, Vol. 29 (1), pp. 89-109.

Panayotou, T., 1993, "Empirical tests and policy analysis of environmental degradation at different stages of economic development", World Employment Programme Research, Working Paper 238, International Labour Office, Geneva.

Pearson, P., 1994, "Energy, externalities and environmental quality: will development cure the ills it creates?", Energy Studies Review, Vol. 6, n.3, pp. 199-215.

Roberts, J.T., and Grimes, P.E., 1997, "Carbon intensity and economic development 1962-91: a brief exploration of the environmental Kuznets curve", World Development, Vol. 25, n. 2, pp. 191-198, Elsevier Science Ltd.

Selden, T.M., and Song, D., 1994, "Environmental quality and development: is there a Kuznets curve for air pollution emissions?", Journal of Environmental Economics and Management n.27, pp. 147-162.

Shafik, N., 1994, "Economic development and environmental quality: an econometric analysis", Oxford Economic Papers, Vol. 46, pp. 757-773.

Stern, D.I., Common, M.S., and Barbier, E.B., 1994, "Economic growth and environmental degradation: a critique of the environmental Kuznets curve", Discussion Paper in Environmental Economics and Environmental Management, n.9409, University of York.

Torras, M., Boyce, J.K., 1998, "Income, inequality, and pollution: a reassessment of the environmental Kuznets curve", Ecological Economics, Vol. 25, pp. 147-160.

Unruh, G.C., Moomaw, W.R., 1998, "An alternative analysis of apparent EKC-type transitions", Ecological Economics, Vol. 25, pp. 221-229.

Vincent, J.R., 1997, "Testing for environmental Kuznets curves within a developing country", Environment and Development Economics, Vol. 2, pp. 417-431, Cambridge University Press.

World Bank, 1992, "World Development Report 1992", Oxford University Press, New York, USA.

Table 1: Cross country studies on EKC indicators

	STUDIES: cross-country										single-country		
Indicator	G 95	GK 94	HLW 97	P 93	SS 94	S 94	TB 98 w/o	TB 98 w/	KDGP98	CRB 97	HES 95	CJM 97	V 97
SO₂	N-shaped ($4107)	N-shaped (4000-5000)		EKC ($2900-3800)	EKC ($10292-10681)	EKC ($3670)	N-shaped	N-shaped	U-shaped or EKC (3)	EKC ($6900)		MD	
SPM	EKC ($15903)			EKC ($4500)	EKC ($9811-10289)	EKC ($3280)				EKC ($7300-18000)		MD	MI
NOₓ	EKC ($18453)			EKC ($5500)	EKC ($11217-12041)					EKC ($14700-17600)		MD	
CO	EKC ($22819)				EKC ($5963-6241)					EKC ($9900)		MD	
CO₂						MI				EKC ($62700)	EKC ($35428)		
CFC										MI in 1986 EKC in 1990 ($12600)			
GHG												MD	
VOC												MD	
AIR TOXICS												MD	
SMOKE		EKC ($6151)					N-shaped	Y not signific					
HEAVY PARTICLES	MD							Y not signific					
AIRBORNE LEAD	inverted-S												
PM₁₀												MD	
DO		MI				MD	MI	N-shaped					
COD	EKC ($10403)	EKC ($7853)											Y not signific
BOD	EKC ($9950)	EKC ($7623)											Y not signific
TC	MI	N-shaped											
FC	EKC ($8193)	EKC ($7955)				N-shaped	MI	Y not signific					
LEAD	MD	first falls, then levels off											
CADMIUM	MD	first falls, then levels off											
MERCURY	N-shaped	N-shaped											
ARSENIC	N-shaped	N-shaped											
NICKEL	inverted-N	N-shaped											
NITRATES		EKC ($10524)								EKC ($15600)			
AMMONIACAL NITROGEN													MI
pH													MI
DEFORESTAT.					EKC ($823-1200)	Y not signific							
MSW						MI				MI			
LACK SANITAT						MD	N-shaped	MI					
LACK SAFE H20						MD	N-shaped	N-shaped					
TOXIC INTENS				EKC ($12790) or MI (1)									
TOT. ENERGY USE										EKC ($34700)			
TRANSPORT ENERGY USE										EKC ($4million)			
TRAFFIC VOLUMES										EKC ($65300)			

INDICATORS LEGEND: SO2=sulphur dioxide. SPM=suspended particulate matters, NOx=nitrogen oxides. CO=carbon monoxide, CO 2=carbon dioxide. CFC=chlorofluorocarbons, GHG=greenhouse gases. VOC=volatile organic carbon, PM10=particulate matters less than 10 microns in diameter. DO=dissolved oxygen, COD=chemical oxygen demand, BOD=biochemical oxygen demand. TC=total coliforms, FC=faecal coliforms, MSW=municipal solid wastes.

AUTHORS LEGEND: G 95=Grossman (1995), GK 95=Grossman and Krueger (1994). HLW 97=Hettige, Lucas and Wheeler (1997). P 93=Panayotou (1993), SS 94=Selden and Song (1994). S 94=Shafik 1994, TB 98w/o=Torras and Boyce (1998) without inequality. TB 98w/=Torras and Boyce (1998) with inequality, KDGP 98=Kaufmann, Davidsdottir, Garnham and Pauly (1998), CRB 97=Cole, Rayner, Bates (1997). HES 95=Holtz-Eakin and Selden (1995), CJM 97=Carson, Jeon and McCubbin (1997). V 97=Vincent (1997).

RESULTS LEGEND: EKC=environmental Kuznets curve. MI=monotonically increasing. MD=monotonically decreasing. Inverted-S shape=environmental degradation first rises, then levels off and finally increases again as income grows. Inverted-N shape=environmental degradation first falls, then rises and finally decreases again as income grows. Y not signific.=income not statistically significant. N-shaped=environmental degradation first rises, then falls and finally rises again. Income level at the turning point in brackets. Minimum and maximum income levels given when several estimates are performed. In CRB 97 values refer instead to turning points without and with transport sector. All values are in 1985US$ unless otherwise specified.

NOTES: (1) The study by de Bruyn et al (1998) is not included among single-country contributions in the table since the authors adopt a different econometric model, taking growth rather than per capita income as explanatory variable. (2) EKC when emission intensity measured per unit GDP. MI when emission intensity measured per unit industrial output. (3) U-shaped with respect to income, EKC with respect to spatial intensity of economic activity (turning point at $6.7million with national spatial intensity, $154million with city-specific spatial intensity).

9. Environmental Resources Valuation as an Institutional Problem

Maurizio Franzini

Introduction

One of the most controversial problems in the environmental field are the criteria for deciding the destination and use of various environmental resources. The criterion proposed by economists which is essentially based on a comparison of costs and benefits has been fiercely criticised for some time, and is often part of a general rejection of the economic approach to environmental questions. For example, more than 10 years ago, Sagoff (1988, p. 1) expressed his radically critical point of view as follows: "the conceptual vocabulary of resource and welfare economics ... which once served to justify social, especially environmental policy, has largely outlived its usefulness and has become a distraction and an important obstacle to progress".

As Pearce-Barbier (2000, p. 7) recently observed, much criticism of the economic approach is prejudiced and based on imperfect understanding of the positions of economists. They go on to say that since there do not seem to be any better alternatives, it is a relief to state that cost-benefit analysis is still "the best game in town" (Pearce- Barbier 2000, p. 83).

This is probably true, but the approach of economists, especially the more traditional one, certainly has weaknesses that could be improved without, as some critics demand, betraying its distinctive features. The main aim of this paper is to discuss two specific problems in the application of the economic criterion of environmental valuation, with a view to correcting them. The first problem is the unreliability of preferences expressed by individuals on whom the calculation of costs and benefits is based. The second concerns the unjustifiably strong influence that richer individuals, who express "heavier" preferences, end up having on final decisions.

In looking at these questions, certain more radical criticisms of the economic method regarding the above questions will be considered, namely the preferences on which valuation is based and whether it is appropriate to make an aggregate of individual preferences, as economists suggest. Many of the criticisms turn out to have poor foundations and to be of little use for formulating better alternatives. However, with adjustments, some enable modifications that could alleviate the above problems. In general, the solutions consist in reinforcing and articulating the institutional context in which decisions are made.

To make individual preferences more reliable, institutions that favour information transfer, exchange of opinions and identification of shared values are required: this cannot occur if valuation is based on attempted market simulation. So that decisional criteria are not excessively conditioned by preferences of the rich, institutions are needed that ensure implementation of decisions and are provided with fail safe mechanisms that prevent exploitation of imperfections in the system to achieve effects in line with the preferences of the rich.

The arguments put forward in the following pages help to explain the reasons and importance of this concept. In the first section, the main features of the economic method and its strengths and weaknesses are described. The reliability of preferences as a basis for valuation is discussed. In the second section we look at extreme positions, such as the claim that decisions regarding the environment should not depend on individual valuation and preferences. We show the limits of this position and how it could lead to undemocratic and non implementable decisions.

In the third section, we consider whether a distinction should be made between citizen and customer, as many sustain, or whether only citizens should be considered as a source of information for economic valuation. The customer/citizen distinction is interesting for reinforcing the reliability of individual preferences, however, as we shall see, it is a complex concept and many of its elements are controversial. We nevertheless obtain useful indications about the best institutional contexts for increasing the reliability of preferences.

In the fourth section, with reference to this problem, we look at a decision-making procedure known as deliberative democracy, that some regard as an alternative to the economic method. We try to show that deliberative democracy cannot be regarded as a method for making decisions in the absence of unanimity, but its reasonable discussion feature can help to increase awareness of individual valuations and thus the reliability of preferences. The institutional changes referred to above should include systematic use of this form of participation and discussion.

In the fifth section we look at the problem of aggregating preferences, the influence of income and the possibility of using income-independent criteria. We go into the question of effectively implementing decisions which do not reflect the preferences of the rich, when enforcement is weak. Once a decision has been made, the richer sector can invest greater resources in lobbying and seeking impunity for non compliance. Our analysis leads us to emphasise that a society wishing to maintain the possibility of making decisions contrary to the preferences of the rich and of effectively implementing them, must have strong and independent institutions.

Environmental valuation not only raises technical problems, but also major institutional ones.

The main features of the economic method of valuation

Individual preferences play a crucial role in the approach used by economists to evaluate environmental resources, known generically as cost-benefit analysis. Preferences are expressed as willingness to pay WTP (or accept WTA), in other words as sums of money that individuals are willing to pay for environmental improvement (or accept as compensation for environmental deterioration). The valuation, and hence the decision, is obtained summing the various WTP or WTA. The three main features of this approach are therefore:

1. individualism;
2. expression of preferences in terms of money (WTP, WTA);
3. aggregation of WTP and WTA by summation.

All these features have been criticised. Some consider the individualistic approach unacceptable; others accept it but take exception to basing valuation and decision on WTP; others object to the criterion of aggregation based on the sum of WTP declared by the various subjects. These criticisms have different weights and implications for the economic method. For example, the objection against aggregation can be accepted without great repercussions, but rejection of individualism is insuperable. Before looking more closely at some of these problems, it is useful to examine the link between method of valuation and market function.

It has been correctly observed that economists use this method "to infer the values individuals would attach to these commodities if market discipline were in place" (Crocker et al. 1998, p. 143). In other words, the aim is to "simulate" market function, with individuals expressing their willingness to pay or accept through prices; if the former exceed the latter, a transaction conforming to the criteria of Pareto efficiency takes place, generally determining an improvement in social well-being. The well known absence of markets for many environmental resources is not sufficient reason, as far as economists are concerned, for not seeking efficiency. This is why they favour a method, that as they see it, leads to decisions similar to those that would be made in a (perfect competition) market.[1]

In examining the strong and weak points of the economic method, it should be borne in mind that the main aim of economists is to observe the criterion of efficiency. This criterion is often interpreted reductively and regarded as equivalent to profit. Although the notion of efficiency is not completely unambiguous,[2] it is certainly not equivalent to profit. According to the original Pareto approach, efficiency was the essential condition for social well-being, not for advantages for a few.

In this paper, social well-being and efficiency as indicated above play an important role. If the importance of the efficiency criterion is recognised, two problems emerge clearly. These problems are worthy of more attention than they have received.

The first problem is the reliability of preferences expressed by individuals or the reliability of their WTP and WTA as a basis for valuation.

Many factors can cause substantial divergence between the preferences of individuals and the actual change in well-being determined by environmental resources that these preferences are intended to represent. If the basis of the calculation is distorted, it is impossible to achieve efficiency.

Sometimes this problem is discussed using the term "true preferences" (Spash 1997); here we consider it more appropriate to refer to reliable preferences because a search for "true preferences" is unlikely to achieve its confused objective.

If preferences of individuals in the environmental field were expressed in the same way as preferences are expressed in any market,

---

[1] There are significant differences between the method of valuation proposed by economists and how the market actually functions. The major difference is that payments are real in the market and only hypothetical in the valuation. The various differences, which are not always adequately considered, may have substantial implications. Some will be examined later.

[2] A recent contribution is by Furubotn (1999).

there would be many reasons for not regarding them as reliable. In fact, they would be subject to information gaps, strategic calculations, incompleteness of preferences and beliefs and many other defects that would make them an unacceptable basis for efficiency-oriented choices.

The second problem regards possible divergence between decisions and effects. A grave defect of most theoretical models for optimal decisions in the environmental field is a tendency to regard decisions as equivalent to the desired effects, whereas significant gaps occur in practice. Since social well-being depends on effects, any divergence should be borne in mind during discussions on the decisions to make in the environmental field. Both these problems, discussed below, have implications of an institutional nature and suggest that the market analogy cannot be taken too far. Valuation has its own specific characters and institutions other than market simulations are required to make preferences more reliable and to ensure that decisions have their intended effects. Economists need to become more aware of the importance of these problems and should cease to take the market as the only reference.

Preferences and economic valuation. Exclusion of individualism?

Much importance is attributed to the reliability of preferences. We now examine the hardest criticism of the fact that the economic approach attributes so much importance to this aspect. The most radical position excludes the possibility of environmental valuation *a priori*. If this possibility is excluded, the subjective judgment of individuals cannot be used as a basis for decisions. Proponents of this view essentially share the idea that nature has an intrinsic, non anthropocentric value and that the rights of other living species, that human valuations always tend to violate, should also be recognised.[3]

In response to such a clear-cut argument, it is as well to ask whether it is in fact possible to dispense with individual preferences and what the consequences of this would be. The first answer is fairly obvious, namely that exclusion of this criteria is equivalent to dispensing with the cardinal principal of any democratic method of decision-making. The economic valuation system can be regarded as democratic for the very reason that it is individualistic, as various authors have underlined.[4] Naturally other methods of decision-making may also be democratic, but none can do

---

[3] Some of these positions are critically reviewed in Farber (1999).

[4] In apparent contradiction with this, cost-benefit analysis is often accused of being undemocratic. This is justified if concretely expressed preferences of individuals are not used as reference, but cost-benefit calculations only vaguely related to preferences.

without some form of individualism.[5]

If this radical approach were taken, there would be the risk of a paternalistic or in any case elitist type solution, in which only the preferences (moral or otherwise) of someone end up counting. Clearly, in human society, one cannot dispense with the preferences of at least some humans when making a decision. This approach, moreover, does not offer any alternative method of reaching a decision, but tends to predetermine the content of the decision, which typically consists in advocating environmental protection in all cases.

On this question, Kerry Turner (1999, p. 17) writes: "These environmentalist positions lead to the advocacy of environmental sustainability standards or constraints, which to some extent obviate the need for the valuation of specific components of the environment. Taking the polar case, some eco-centrics seem to be arguing that all environmental resources should be conserved regardless of the costs of such strategy, i.e. that environmental assets are infinitely valuable and the environmental standards are absolute".

The assumption of this position, that democratic methods of decision-making always lead to decisions contrary to environmental protection, is extreme and difficult to agree with. There is no reason to regard democracy and environment as always being in conflict, especially if preferences are based on adequate information and not characterised by other distortions that limit their reliability.

From our point of view, the interesting point is another, namely the likelihood that if such radical objectives were adopted officially, they could be unattainable for the very reason that they do not consider the costs involved.

Since the willingness to sustain costs is inevitably in the sphere of individual preferences, the problem can be seen as the consequence of not considering individual preferences. To clarify this point, a concrete example is useful.

The Endangered Species Act approved by the United States in 1973 is based on the conviction that the value of endangered species is incalculable and substantially that all species in danger must be saved. The species to protect were listed as more than 1100 (either threatened or endangered) but the funds available were insufficient to protect them all (Brown Jr. -

---

[5] For example, individuals aware of their ignorance may delegate their decisions to "experts" (Sunstein 1999). However for individuals to freely seek the advice of experts is not the same as for experts to impose their decisions. As we shall see, the best solution could be one involving institutions that inform individuals so that they can make well informed valuations.

Shogren 1998). This lack of funds can be regarded as proof of the fact that the society is not ready to sustain the costs that conservation involves. The consequence is that only some species are effectively protected and which ones they will be is decided by the bureaucrats who manage the funds. This seems contrary to the most elementary ethical principles.

The example illustrates how limited willingness to sustain the costs of conservation is a serious obstacle in the decisive phase of implementation of decisions. This obstacle cannot be overcome by paternalistic decision-making but requires more complex intervention of an institutional type. The problem of the relation between decisions and implementation is discussed in more detail below. However, it is important to underline that how decisions are made may play a major role in determining how far the effects fall short of the decisions.

Interests and values. From consumers to citizens?

The use of preferences as a basis for evaluating environmental resources has been the subject of another severe, though less radical, criticism. It was formulated with particular conviction by Sagoff (1988) and regards the distinction between individuals as consumers and individuals as citizens.[6]

The former are said to be motivated by a rather extreme form of self-interest, whereas the latter also consider the interests of others in their decisions and are guided mainly by values. Thus expressed, the distinction between citizens and consumers is not entirely convincing. Among the many critical positions, let us look at that of Kerry Turner (1999, p. 30): "Is it really the case that the consumer/citizen distinction is absolutely clear-cut? Some experimental evidence indicates that even in the context of market transactions some individuals refuse to trade or accept rewards if they perceive the operating conditions are 'unfair'".

This quote suggests that the motivations of consumers may also be guided by moral values and not only by self-interest. It may be added that the most widely used method for determining WTP, known as Contingent Valuation, is appreciated by economists because it seems also to reflect valuations that do not depend solely on the user value of the environmental resource but also on its existence value.[7] If it is recognised that individuals

---

[6] The distinction has often been proposed by economists in this and other contexts. See for example Marglin (1963).

[7] Some sustain that even methods of determining WTP based on actual market behaviour (so-called indirect methods) may reflect existence value if the utility functions have the appropriate characteristics. See, for example, Larson (1994).

can attribute existence value to environmental resources, and the trouble is taken to determine this value, the accusation that valuation is based on consumers as understood by Sagoff can hardly be made.

This does not mean that the criticisms do not contain elements worthy of consideration. However, the distinction between citizen and consumer seems to involve too many elements which obscure the true problem, namely the relationship between individual valuations and methods to determine them.

Generalising, we can say that the question regards the relation between preferences and the institutions where they are expressed. Before examining this in more detail, let us look at the less convincing ideas underlying the citizen/consumer distinction.

There is a common unjustified prejudice against the WTP as a method of measuring preferences. The expression of preferences in monetary terms is regarded as morally unacceptable because it evokes the idea that individuals are willing to exchange the environment for vulgar monetary compensation.

The problem should however be examined from a different perspective, namely what sacrifice, in terms of willingness to give up other sources of well-being, is the individual prepared to make to safeguard environmental resources? If a relation exists between value and sacrifice, this question seems legitimate, and the use of money as a unit is justified by ease of calculation. If the WTP were expressed in terms of bread it would probably raise less criticism because it would be clear that the method is based on the idea that for each of us, the environment has a value equal to the sacrifice we are willing to make in terms of other goods or any other element, even non material, that affects individual well-being.

Of course some may not be willing to sacrifice anything for the environment, in which case the responsibility of using money as a unit is not clear. Others expect everyone to be willing to sacrifice anything for the environment. As much as one may agree with this, one has to admit that other moral values can legitimately be pursued by individuals. The mere existence of these values makes the idea of unlimited sacrifices to safeguard the environment, ethically unsustainable, besides probably being unattainable in practice.

Behind the consumer/citizen distinction, there is probably the idea that the mere request to express one's preferences in money corrupts the valuation.

Were this true, the position would hardly be convincing. However, as we have said, other interesting elements of the relation between institutions and preferences are revealed by this distinction.

It is important how preferences or WTP are expressed. Much of the literature on Contingent Valuation is concerned with the sensitivity of preferences in relation to the way in which they are measured. Contrary to frequent claims, many economists are therefore aware of this difficulty.

The sensitivity of individual valuations is influenced by many factors in relation to the context in which they are expressed. These include:

1. lack of information on the consequences of the actions to evaluate;
2. the possibility of strategic behaviour and free riding;
4. incompleteness of preferences and beliefs;
5. influence of irrational elements.

Different contexts of expression of preferences affect each of these elements and end up affecting the valuations expressed by individuals. In choosing the institutions to use, these elements should be borne in mind.

Those that provide the best information, reduce the risk of strategic behaviour and incompleteness of preferences, and are affected least by irrational decisions should be selected.

To choose well requires knowledge that is not available, especially with respect to certain problems. However, what we already know is probably enough to ensure some progress in the direction of an institutional setting that improves the reliability of preferences expressed by individuals. In this perspective, reference to the citizen could be useful, not because he is more moral than the consumer, but rather because he expresses his valuations in an institutional context that reduces the information shortage, limits strategic behaviour and is generally less subject to factors that distort WTP.[8]

Alternatives to the economic method: deliberative democracy

Now let us ask what alternatives to the conventional economic method can offer to improve the reliability of preferences. If we exclude the paternalistic alternative, not many others have been proposed. This is one

---

[8] The following statement of Sunstein (1999, p. 42) suggests an interpretation similar to the present. It is related to the second factor in our list: "aggregating private willingness to pay can replicate various collective action problems faced in the private domain; people may be willing to pay more simply because they know that other people are contributing as well. If this is so, it makes no sense to base policy on private willingness to pay, where the collective action problem arises." Institutions that favour direct contact between individuals to reduce the risk of free-riding lead to higher and more reliable individual valuations.

reason why some economists claim that cost-benefit analysis is still "the best game in town" (Pearce-Barbier 2000, p. 83).

Sustainers of the consumer/citizen distinction regard deliberative democracy as a true alternative to the economic method and advocate it for decisions on environmental questions. This form of democracy, as distinct from representative democracy, has recently found renewed interest with scholars (see for example Elster 1998).

Opinions on the features of deliberative democracy are rather varied and not always mutually reconcilable. In some cases the accent is placed on the results that can be obtained, in others on the procedure itself. The latter can be of interest. In this regard, Elster (1998, p. 8) writes: "the notion includes collective decision making with the participation of all who will be affected by the decision or their representative: this is the democratic part. Also, (....) it includes decision making by means of arguments offered *by* and *to* participants who are committed to the values of rationality and impartiality: this is the deliberative part".

To express a mediated judgment on the contribution that deliberative democracy can make to overcoming the problems of environmental decisions, it is useful to make a distinction. On one hand, we have the question of the reliability of preferences, on the other the problem of aggregation of preferences and actually reaching a decision. In the debate on deliberative democracy, a widespread idea is that ample, reasonable collective discussion favours attainment of unanimity, dispensing with the arduous task of deciding when there are conflicting opinions. This is not impossible, but the idea that deliberative democracy always leads to unanimity has little real basis.

Johnson (1998) made many relevant observations in his critical evaluation of deliberative democracy. He sustained that aggregation through vote (the problems of which have been well known since the famous work of Arrow (1951)) cannot be replaced by deliberative democracy. However, he considers it possible that the "reasonable discussion" feature of deliberative democracy may reduce the range of preferences, facilitating aggregation.[9]

Irrespective of its capacity to make individual preferences more homogeneous, the "reasonable discussion" feature of deliberative democracy could incease the reliability of valuations expressed by

---

[9] He sustains that single-peaked preferences could be formed. It is well known that these are sufficient to overcome the Arrow impossibility theorem (Johnson, 1998, p. 177). In the discussion typical of deliberative democracy, negative phenomena promoting conflict rather than limiting it may also occur. This is stated by Johnson (1998, p. 167), and Elster (1998) considers the phenomenon of intellectual wars that create distance between the various positions, rather than reconciling them.

individuals. Exchange of information, direct comparison of the various positions, and the relations that form between the various subjects can have a positive effect on factors limiting the reliability of preferences. In particular, it can reduce the negative weight of the lack of information and by favouring a sense of belonging to a community, the inclination to free ride.

In the sense that it helps to increase the reliability of preferences, deliberative democracy would be a useful component of an acceptable decision-making procedure.[10] Many aspects of its use naturally need further clarification. Contrary to what many sustain, however, it is not a complete decision-making procedure. Other elements are necessary and other problems have to be solved. A major one is how to aggregate inevitably diverging preferences.

Decisions and their effects. The importance of methods of aggregation of preferences

In the choice of method of aggregation, the main question regards the "weights" to attribute to each individual. In the traditional voting system, individuals all have the same weight, but if we make aggregates of monetary quantities such as the WTP and WTA, individuals have different weights.

There are various reasons for preferring either of these methods. Before examining them, it is worth saying that the two methods do not necessarily lead to different decisions, and thus in some cases, it is irrelevant which is chosen.[11]

One of the essential differences between the two methods is the different profile that they attribute to the intensity of preferences. Using the WTP means that collective decisions will depend not only on preferences of individuals but also on their intensity. Even if we consider it positive to have a decision-making system sensitive to the intensity of

---

[10] Here again there are dangers. The strongest group may manipulate the preferences of others to their advantage. Institutions should guard against this phenomenon, which is not exclusive to the present context, but possible in many others.

[11] Identification of preferences by one of the methods preferred by economists, namely Contingent Valuation, may often occur in the form of a genuine referendum. Individuals are simply asked whether they would be willing to pay a certain sum. In this case, the differences in terms of expression of vote between the two systems are practically zero. This does not mean, however, that information thus obtained is used as it is in a political voting system .

preferences, it is well to ask whether mere aggregation of WTP leads to unjustifiably distorted results.

The main problem is the influence of the distribution of incomes. WTP is affected by income because of income effects and because lower incomes effectively limit the willingness to pay of many persons. In general, therefore, the results of the decision-making process are affected by changes in income distribution and this raises the delicate, difficult problem of establishing the income distribution to make an ideal basis for the expression of individual evaluations. An attractive idea is to use an egalitarian income distribution. In a world in which significant inequality exists between incomes, a WTP expressed in this way would have to be adjusted to purge it of the influence of this inequality. The aim is to try to establish what the WTP of individuals would be if their incomes were all the same.[12]

Eliminating the weight of income on individual WTP means trying to ensure that the preferences of the rich do not count more than those of the poor, without, of course, going as far as the principle "one head, one vote" typical of political voting. The adjustment should not eliminate all the influence of the intensity of preferences; the aim is only to make the various WTP express the different structure of preferences and not the different availability of income as well.

As we said, this criterion of aggregation is attractive in terms of equality, although it is not true, as sometimes claimed, that it necessarily leads to decisions more favourable for the environment. Indeed, if willingness to spend for the environment increases significantly with income, as some claim, the effect of an even theoretical levelling of incomes could reduce the overall willingness to pay for the environment.

Irrespective of this possibility, however, another argument should be remembered in deciding the best method of aggregation of preferences, namely the possibility that the effective results are significantly different from the intention of the decisions. This problem presents in different ways according to the overall functioning of the institutions and this particular institutional property is important for our discussion.

The question is of the same kind as the one illustrated with the case of the Endangered Species Act. Then, we showed that a system that ignores the effective willingness of individuals to pay may lead to decisions which are unlikely to achieve their aims in practice. Now we can generalise the question to consider the functioning of institutions that implement

---

[12]   There are many ways of making this correction (see Pearce-Barbier 2000) though none are perfect.

decisions and those that solve the complex problems of monitoring and enforcement.

When decisions involving benefits for some individuals and costs for others are made, there is always a substantial risk that those damaged take steps to prevent the decisions from being implemented, or to avoid paying the costs requested. In both cases, a gap is created between decision and effects. The size of this gap depends on whether the institutions charged with putting decisions into practice can be influenced (for example by lobbying) and on the weakness of the action of institutions charged with making sure costs are paid, in whatever way they manifest.

With regard to the latter, the problems of enforcement of environmental regulations are great and have been the subject of many studies. These studies generally have not linked non compliance with the decision made, in other words they do not evaluate the severity of the problems of enforcement in relation to the gap between the decision and the effective willingness of the collectivity to pay. However it seems possible to say that when the institutions of enforcement and implementation are imperfect, the less the decisions reflect the effective willingness of the collectivity to pay, the greater the gap between effects and decisions. If those who pay costs were appropriately compensated by those who benefit, as occurs in the market with the payment of prices, then the problem of enforcement would be greatly reduced or even completely eliminated.

The above considerations should play a role in deciding what system of aggregation of preferences is best. It is most unsatisfactory, even morally, to make decisions that are systematically violated. If results and not decisions are what count here, as in the field of social well-being and efficiency, it is better to opt for decisions that are less in line with abstract egalitarian criteria but are more easily put into practice, than for ethically impeccable criteria that cannot be implemented. This option can be taken without violating any ethical principle.

We can therefore say that elimination of the influence of income may only seem to be a more ethical criterion than the conventional expression of WTP. For example, if institutions are weak and the decision penalises the rich, the latter will exert pressure that makes the decision ineffective. More detailed analysis, which is not appropriate here, is necessary for a more precise valuation.

There were two reasons for raising these problems in this paper. The first was to emphasise the need to consider problems of implementation when defining decision-making criteria. The other was to show the essential role played by institutions in enabling a society to effectively achieve results that may or may not be favoured by the rich. If the institutions are weak, the preferences of the rich will prevail whether or not the decision-making criteria are in their favour. A society intending to act ethically should not run such risks.

It is therefore essential to reinforce the relevant institutions in order to have greater freedom of choice in decision-making criteria, and if desired, to get away from the economic method. In its conventional form, this method gives more weight to richer individuals, reducing the possibility of having to face the problem of enforcing a decision that damages them. Much theoretical work is needed in this field. Economists need to make a contribution here, overcoming, as they have in other fields, the inclination to regard the market as the institution of reference and the only one that can ensure efficient results. The progress made in economic analysis of institutions could be useful for this task.

Conclusions

In this chapter, an attempt was made to sustain that it is not appropriate and probably not even possible to adopt decision-making criteria in the environmental field that are not based on preferences of individuals. This would hold even if we considered the problems arising due to the fact that future generations cannot express their preferences. It is inevitable that future generations are in the hands of the present ones, so that the preferences of at least some components of the latter end up being decisive, even when they do not seem to be. This is why the position of those who advocate dispensing with any democratic basis of decision-making is unacceptable on the grounds that it favours representation of the interests of future generations.

Inevitable reference to individual preferences does not, however, imply that they are always reliable as a basis for calculations. As we have said, we need to find the best institutional context for the expression of reliable preferences. Deliberative democracy has elements that could be useful to build these contexts. Commitment to finding a way to make preferences reliable does not mean that they have to be invariant with respect to the institutions and social contexts in general. Often conditions of this kind are made for considering preferences acceptable as a basis for calculations.

In our opinion, such conditions are too restrictive, almost impossible to meet and probably unnecessary. Individuals are sensitive to the institutions under which they work, but this does not mean that the diversity of preferences expressed in different cases demonstrates that all preferences are unreliable. This manner of stating the problem shows that the initial assumption is perfect rationality of the individual. If this assumption fails, institutions need to be assessed and those that favour more conscious and less distorted expression of preferences should be chosen.

Many economists have long abandoned the assumption of full rationality, and as a consequence have discovered the importance of institutions other than the market and the need to choose between various imperfect solutions. The time seems to have come to apply a similar approach to the problems of environmental valuation. A further positive consequence would be that the market would cease to be the only institution of reference, as has happened in the field of economic analysis of institutions.

Major institutional problems also arise with respect to the other questions examined in this paper: the possibility of making decisions less dependent on income without rendering decisions ineffective. Here, institutions charged with implementation and enforcement play a crucial role. If they are weak, it becomes almost impossible to implement decisions that do not match the preferences of the rich. In a just society, this possibility should not be excluded. In this case too, the contribution of economic analysis of institutions can be important. It is nevertheless unquestionable that the chances of achieving this aim are very small if institutions become increasingly subordinate to the market. If this should happen, however, the method of valuation proposed by economists could hardly be blamed.

REFERENCES

Arrow, K. J. (1951), Social Choice and Individual Values, New Haven, Yale University Press.

Brown jr., G. M.; Shogren, J. F. (1998), Economics of the Endangered Species Act, Journal of Economic Perspectives, 12, n. 3, pp. 3-20.

Crocker, T. D.; Shogren, J. F.; Turner, P. R. (1998), Incomplete beliefs and nonmarket valuation, Resource and Energy Economics, 20, n. 2, pp. 139-162.

Elster, (1998) Introduction, in Deliberative Democracy, edited by J. Elster, Cambridge, Cambridge University Press.

Farber, D. A. (1999), Eco-pragmatism, Chicago,The University of Chicago Press.

Furubotn (1999), Economic Efficiency in a World of Frictions, European Journal of Law and Economics, pp. 179-197.

Johnson, (1998) Arguing for Deliberation: Some Skeptical Considerations, in Deliberative Democracy, edited by J. Elster, Cambridge, Cambridge University Press.

Larson, D.M. (1994), On Measuring Existence Value, Land Economics, pp. 377-388.

Marglin, S. (1963), The Social Rate of Discount and the Optimal Rate of Investment, Quarterly Journal of Economics, pp. 95-111.

Pearce, D.; Barbier E. B. (2000), Blueprint for a sustainable economy, London, Earthscan.

Randall, A. (1999), Making the Environment Count. Selected Essays of Alan Randall, Chelthenam,Edward Elgar.

Sagoff, M. (1988), The Economy of the Earth. Philosophy, Law, and the Environment, Cambridge,Cambridge University Press.

Spash, C. (1997), Environmental Management Without Envirnmental Valuation?, in Valuing Nature, edited by J. Foster, London and New York, Routledge.

Sunstein, C. R. (1999), Cognition and Cost-Benefit Analysis, Chicago Working Papers in Law and Economics (2nd series) n. 85, october.

Turner, K. R. (1999), The place of economic values in environmental valuation, in Valuing environmental preferences, edited by Ian J. Bateman and Kenneth G. Willis, Oxford, Oxford University Press.

Printed and bound by CPI Group (UK) Ltd, Croydon, CR0 4YY

22/10/2024

01777626-0007